Market Based Fisheries Management

Market Based Fisheries Management

Bruno Augusto Amato Borges

Delve
PUBLISHING
www.delvepublishing.com

Market Based Fisheries Management

Bruno Augusto Amato Borges

Delve Publishing
224 Shoreacres Road
Burlington, ON L7L 2H2
Canada
www.delvepublishing.com
Email: orders@arclereducation.com

ISBN: 978-1-77407-721-4 (Hardcover)

ABOUT THE AUTHOR

Bruno Augusto is Aquaculture Engineer and Master on Aquaculture and Fishery Recources from Federal University of Santa Catarina (UFSC). His focus is on writing projects and development of new technologies in aquaculture. He is currently project service provider, consultant and develops research on Biofloc Technology (BFT) cultivation for fish and shrimp.

TABLE OF CONTENTS

LIST OF GLOSSARY

A

Aborigines – Person, animal, or plant that has been in a country or region from earliest times.

Accidentally – By chance; inadvertently.

Accustomed – Make someone or something accept (something) as normal or usual.

Aggravating – Making a problem or offence worse or more serious.

Aggravation – The state of becoming worse or more serious; exacerbation.

Aggregating – Collect (related items of content) so as to display or link to them.

Allocation – The action or process of allocating or sharing out something.

Ambit – The scope, extent, or bounds of something.

Anchovies – Is a small, common forage fish of the family Engraulidae.

Anglers – Plural for an angler fish, *Lophius piscatorius*.

Anguish – Severe mental or physical pain or suffering.

Arduous – Involving or requiring strenuous effort; difficult and tiring.

B

Bait – Food placed on a hook or in a net, trap.

Biodiversity – Is typically a measure of variation at the genetic, species, and ecosystem level.

Brackish water – Is water having more salinity than freshwater, but not as much as seawater.

Buoy – An anchored float serving as a navigation mark, to show reefs or other hazards, or for mooring.

C

Calibration – Calibration is the comparison of measurement values delivered by a device under test with those of a calibration standard of known accuracy.

Certainty – Firm conviction that something is the case.

Chaos – Complete disorder and confusion.

Chronological – Arranged in the order of time.

Coastal Area – Is the area where land meets the sea or ocean, or a line that forms the boundary between the land and the ocean or a lake.

Conceal – Not allow to be seen; hide.

Conformity – Behavior in accordance with socially accepted conventions.

Cons – Persuade (someone) to do or believe something by lying to them.

Consignment – A batch of goods destined for or delivered to someone.

D

Defied – Appear to be challenging (someone) to do or prove something.

Demersal fish – Are bottom feeders.

Devour – Eat hungrily or quickly.

Dipping – Falling steeply.

Discretion – The quality of behaving or speaking in such a way as to avoid causing offence or revealing confidential information.

Distrust – The feeling that someone or something cannot be relied upon.

Dread – Very unpleasant, serious, or shocking.

E

Efficacy – Ability to produce a desired effect under ideal testing conditions.

Emission – The production and discharge of something, especially gas or radiation.

Enticement – The desire to do something, especially something wrong or unwise.

Estuaries – Is a partially enclosed coastal body of brackish water with one or more rivers or streams flowing into it, and with a free connection to the open sea.

Evaded – Escape or avoid (someone or something), especially by guile or trickery.

Exploited – Make use of (a situation) in a way considered unfair or underhand.

Extrapolating – Is a type of estimation, beyond the original observation range, the value of a variable on the basis of its relationship with another variable.

F

Fierce – Having or displaying an intense or ferocious aggressiveness.

Fleet – A group of ships sailing together, engaged in the same activity, or under the same ownership.

Flouted – Openly disregard (a rule, law, or convention).

Fyke – A bag net for catching fish.

G

Gangling – Tall, thin, and awkward in movements or bearing.

H

Habitats – A habitat is the type of natural environment in which a particular species of organism lives.

Harvested – Focused, centered, achieving a state of shared mysticism.

Haul pull or drag with effort or force.

Heirs – A person legally entitled to the property or rank of another on that person's death.

Herring – Herring are forage fish, mostly belonging to the family Clupeidae.

Horsepower – Is a unit of measurement of power, or the rate at which work is done, usually in reference to the output of engines or motors.

Hostility – Unfriendliness or opposition.

I

Impressionist – A person who follows or adheres to the theories, methods, and practices of impressionism, especially in the fields of painting, music, or literature.

Indisputable – Not open to question.

Intrusion – The act of disrupting something.

J

Jostle – To bump or push roughly.

Jurisdiction – Is the practical authority granted to a legal body to administer justice within a defined field of responsibility, e.g., Michigan tax law.

Juvenile – Is an individual organism that has not yet reached its adult form, sexual maturity or size.

L

Legislation – Is law which has been promulgated by a legislature or other governing body or the process of making it.

Lobbyists – Is the act of attempting to influence the actions, policies, or decisions of officials, most often legislators or members of regulatory agencies.

Longlines – Is a commercial fishing technique which uses a long line, called the main line, with baited hooks attached at intervals by means of branch lines called snoods (or gangions).

Lure – tempt (a person or animal) to do something or to go somewhere, especially by offering some form of reward.

M

Maneuverability – Able to be maneuvered easily while in motion.

Methodologies – A system of methods used in a particular area of study or activity.

Modernization – The process of adapting something to modern needs or habits.

Mortality Rate – Is a measure of the number of deaths in a particular population, scaled to the size of that population, per unit of time.

Mortgage – A thing that is borrowed, especially a sum of money that is expected to be paid back with interest.

Multibeam – Is a type of sonar that is used to map the seabed.

Multinomial – A mathematical expression that consists of the sum of several terms.

N

Nearshore – Of, relating to, or situated in the waters near the shore.

O

Obstruction – The action of obstructing or the state of being obstructed.

Oceanographic – Is the study of the physical and biological aspects of the ocean.

Overfishing – Is the removal of a species of fish from a body of water at a rate that the species cannot replenish in time, resulting in those species either becoming depleted or very under populated in that given area.

P

Paradigm – A set of beliefs or a way of doing, or thinking about, something.

Pelagic – Pelagic zone consists of the water column of the open ocean, and can be further divided into regions by depth.

Perpetual – Never ending or changing.

Phytosanitary – Is used to attest that consignments meet phytosanitary import requirements and is undertaken by an NPPO.

Plague – Any contagious disease that spreads rapidly and kills many people.

Predatory – Living by killing and eating other animals.

R

Recreational – For, or relating to, recreation.

Regenerating – To re-create, reconstitute, or make over, especially in a better form or condition.

Regime – A system or ordered way of doing things.

Rivalrous – Prone to or subject to rivalry.

Robustness – The ability to withstand or overcome adverse conditions or rigorous testing.

Rudimentary – Involving or limited to basic principles.

S

Safeguarding – Protect from harm or damage with an appropriate measure.

Sanitary – Relating to the conditions that affect hygiene and health, especially the supply of sewage facilities and clean drinking water.

Seines – Catch or try to catch fish, typically by using a net or hook and line.

Sine-qua-non – An essential condition; a thing that is absolutely necessary.

Snagging – Catch or tear (something) on a sharp projection.

Socio-economy – Is the social science that studies how economic activity affects and is shaped by social processes.

Staging – The method of presenting a play or other dramatic performance.

Staked – Mark an area with stakes so as to claim ownership of it.

Stimulating – Is the encouragement of development or the cause of activity generally.

Subsidized – Having part of the cost of production paid in order to keep the selling price low.

Suspended – Temporarily prevent from continuing or being in force or effect.

T

Tilapia – Are mainly freshwater fish inhabiting shallow streams, ponds, rivers, and lakes, and less commonly found living in brackish water.

Towed – Pull along with a rope, chain, or tow bar.

Transplantation – He process of taking an organ or living tissue and implanting it in another part of the body or in another body.

Trawlers – Is a commercial fishing vessel designed to operate fishing trawls.

U

Unimpeachable – Of such a high standard of honesty and moral goodness that it cannot be doubted or criticized.

Upwelling – A sudden occurrence of or increase in a phenomenon, feeling or emotion.

W

Wrangling – Engagement in a long, complicated dispute or argument.

LIST OF FIGURES

LIST OF TABLES

LIST OF ABBREVIATIONS

AHRs	annual harvesting rights
ATFS	American tree farm system
BALTFISH	Baltic Sea Fisheries Forum
CFPs	certified forest products
CONSSO	Conference on North Sea Senior Officials
CPUE	catch per unit effort
CSA	Canadian Standards Association
CTBT	Committee on Technical Barriers to Trade
CTE	Committee on Trade and Environment
DGPS	differential global positioning system
DNA	deoxyribose nucleic acid
EEZ	exclusive economic zone
EFH	essential fish habitat
EIA	environmental impact assessment
EU	European Union
FADs	fish aggregation devices
FAO	Food and Agricultural Organization
FLO	Fair-Trade Labeling Organizations International
FMPs	fishery management plans
FSC	Forest Stewardship Council
GDP	gross domestic product
GEN	global eco-labeling network
GHGs	greenhouse gases
HELCOM	The Helsinki Commission for the Helsinki Convention on the Protection of the Marine environment in the Baltic Sea Area
IFOAM	International Federation of Organic Agriculture Movements
IFQ	individual fishing quotas
IOC	Intergovernmental Oceanographic Commission

ISEAL	International Social and Environmental Accreditation and Labeling Alliance
ISO	International Organization for Standardization
ITQs	individual transferable quotas
IUCN	International Union for Conservation of Nature
IUU	illegal, unreported, and unregulated
LAT	lowest astrological tide
LCA	life cycle assessments
LOS	law of the sea
MBES	multibeam echosounder
MPAs	marine protected areas
MSC	marine stewardship council
MSFCMA	Magnuson-Stevens Fishery Conservation and Management Act
MSFD	marine strategy framework directive
MSP	marine spatial planning
MSY	maximum sustainable yield
NEI	not elsewhere included
NGO	Non-Governmental Organization
nprPPMs	non-product related process and production methods
OECD	Organization for Economic Co-Operation and Development
OSPAR	The OSPAR Convention for the Protection of the Marine Environment of the North-East Atlantic
PEFC	Programme for Endorsement of Forest Certification Schemes
PPMs	process and production methods
PSAT	pop-up satellite archival tag
PVC	polyvinyl chloride
RNA	ribonucleic acid
SEA	strategic environmental assessment
SFA	sustainable fisheries act
SFI	sustainable forestry initiative
SPS	sanitary and phytosanitary measures
SSF	small-scale fisheries
TAC	total allowable catch
TACC	total allowable commercial catch

TBT	technical barriers to trade
TRUFs	territorial use rights in fisheries
UN	United Nations
UNCLOS	United Nations Convention on the Law of the Sea
UNEP	United Nations Environment Programme
UNESCO	UN Educational, Scientific, and Cultural Organization
VIIRS	visible infrared imaging radiometer suite
VMS	vessel monitoring system
WTO	World Trade Organization

PREFACE

Fisheries form an integral part of the food culture and business. Fish culture is a crucial contributor to the food requirements and demands of the people across various countries, especially in the Asia Pacific region. Fish and fish-based products from the staple diet of most of the people in these regions. This makes the management of the fisheries significant in this region.

The fisheries form a huge chunk of the market. There are other purposes also for which the fisheries are carried out, like the various recreational and entertainment purposes. However, the food-based market is the biggest market in terms of the activities related to the fisheries.

The management of fisheries for a certain market requires a lot of inputs and contributions. There are several working principles for the fisheries management and considerations that are in place for the said purpose. The fisheries management incorporates some specific techniques, which are to be undertaken for the desired results and benefits. These aspects of fisheries management have been informed about in depth in this book.

Moreover, there are several rights in the management and ownership of the areas and fields meant for the fisheries purposes. The book discusses those rights with the readers and also makes them aware of the various issues that are involved with the manner in which those rights are exercised.

The readers are further informed about the current scenario of the fishing industry and the various kinds of problems that are being faced by the marine fisheries. There is a focus on the production of the capture fisheries and the limits set to them globally, along with the various kinds of problems associated with them.

The book discusses the significance of eco-labels in the packaging of the fish-based products for the purpose of selling in the market. It also draws the attention of the readers to the process of eco-labeling in the developing countries and makes them aware about the importance of such a process.

There are several activities that go on in the fisheries industry and the production of the various goods are carried out in several ways. The book brings the attention of the readers to the different kinds of fishing gear, the types of traps used to catch the fishes, and the various modes of production in the fisheries industry.

As the production of the fish-based goods increases, so does the pressure on the fishes in the marine ecosystem. It becomes important for the management to plan for the conservation of the fishes and keep into account the marine biodiversity before they head out to catch the fishes for the production of the goods. Hence the readers are informed about the challenges that lie in the fisheries industry and the difficulties that may arise in the fisheries management.

The artificial reefs provide the ecosystem in which the fishes feel comfortable and can breed by laying eggs. The fisheries industry invests a lot of amount in the artificial reefs so that they can be utilized for the desired results. The artificial fisheries have been elaborated upon, in the book.

The market arising from the fisheries and its development has a huge potential to feed and serve the purposes of the people around the world. The future holds good for the fisheries industry and hence, the management of the industry assumes a crucial role. The various possibilities in the fisheries industry have been discussed and focused on, in detail.

The book aims to provide the readers with a thorough knowledge on the fisheries industry and the management techniques that are helpful and employable in shaping this industry. I hope that the book is instrumental in achieving the objectives of educating the readers and spreading knowledge and information to the minds of the people.

1

Introduction to Market-Based Fisheries Management

CONTENTS

This chapter which is named 'Introduction to Market-Based Fisheries Management' provides the information regarding the market-based fisheries management and its working principle. In the first half, the chapter tells about the fishery manager who are the managers of fishery and who are responsible for the decisions taken considering the management of fishery.

It provides readers with the insights of goals and objectives that should be set while managing the fisheries and taking decisions regarding the management. in the second half, the chapter focuses on the fisheries management considerations such as biological considerations, ecological considerations and much more. This chapter also explains how to design a management plan and why it is important.

1.1. INTRODUCTION

Fisheries management approaches fisheries science so as to find the ways in order to protect fishery resources.Fisheries management can be referred to as 'the integrated process of information which gathers, analyses, plans, consults, makes decision, allocates resources as well as formulates and implements, with enforcement as per the requirements, regulations or rules. These rules and regulations govern fisheries activities so as to make sure the continued productivity of the resources and the accomplishment of some other fisheries objectives.'

Fish as well as fisheries are said to be an important part of most of the societies and further make important contributions that is made to the economic as well as social health and well - being in a number of different countries and areas.

A system of appropriate, objectives which are scientifically based and rules for management are all included in fisheries management. Strategies as well as tools for implementing such rules should also exist, along with a system so as to enforce the rules and further perform the monitoring the results.

Social as well as economic benefits can be increased by implementing a successful fisheries management system while maintaining the sustainable yields and protecting the fishery resources.

The management of the marine fisheries of the world tends to affect both large as well as small-scale. Its effectiveness is important for the food provision and livelihoods which depend upon fish and seafood, and to other aspects of ocean health.

Typically for large-scale fisheries, management is the responsibility of federal, state or town government. Considering the small-scale fisheries, management can be done by the collaboration at more local levels, which includes communities, tribes or even families, along with the relevant government entities.

The effectiveness of the large-scale fisheries management is used as a resilience component for food provision (fisheries), natural products (such as fish oil), sense of place (such as ionic species) and biodiversity (such as species and habitats).

The effectiveness of the small-scale fisheries management is used as a status component for the artisanal fishing opportunity and a resilience component for the coastal protection (such as coral reefs layer) and biodiversity (species).

1.2. WORKING PRINCIPLES OF FISHERIES MANAGEMENT

Management is considered to be an important part of any development process. There should be an effective management plan which can be included in the development strategy. This seeks to establish the best kind of relationship.

Furthermore, it suggests that in the initial phase when there is increase in the effort then the yield also increases. Later on, when the increase in the effort slows down in comparison to that of effort, there is continuous decrease in the yield per unit area.

Until the management intervenes, the fisheries will further progress and reach the maximum yield. Further, it will progress and reach at a point where the yield is equal to the cost of the fishery. It is the only situation when it is important to have management and so, compensation is provided by the government in the form of aids and subsidies and further, it makes attempt so as to continue profitability.

When all the possibilities are exhausted, the fishery will not achieve the point and cannot further go beyond the point when only running cost such as the cost for the crew, fuel is met with no fish yield and government aids. The fishery arises further at this stage of decision, it is clear that the crisis cannot be solved until unless the use of strict action is done.

Figure 1.1.*Commercial fisheries in the year 19463.*

Source: Image by Flickr.

In early development phase, it is quite easier to control in comparison to reducing the exploitation when the crisis arises. In India, marine fisheries remained in its pre-developed phase will the year 1962 (that is referred to as pre mechanization period; annual production: less than 0.8 m t).

Then there is a prolonged growth phase till the year 1988 (referred to as mechanization period; when there is increase in the number as well as efficiency of the fishing vessels; annual production: between 0.8 to 1.8 m t); which is followed by the entirely exploited phase, which further lasted for 15 years till the year 2003 (exploitation of unexploited coastal areas as well as there is increase in the effort; annual production: between 1.8 to 2.6 m t).

The Peruvian anchovy fisheries was considered to be one of the largest fisheries during the boom years. In the year 1953, the first fish meal was developed. In the further nine years, Peru became number one in the fishing nation of the world. Approximately 1700 purse seines exploited around seven months in the fishing season. In the year 1972, sustainable yield was estimated approximately 9.5 million tones.

In the year 1969, the collapse of Norwegian as well as Icelandic herring fisheries took place, because of which Peru got more cash and further it allowed to harvest around 12.4 million tones. Further, in the following year 1972, it was around 10.5 million tones.

In this year, El-Nino, and overfishing for a long period resulted in collapse, which have never been recovered. Over the time scale, the development of fishery can be categorized as:

1. Predevelopment phase;
2. Growth phase;
3. Full exploitation phase;
4. Overexploitation phase;
5. Collapse phase; and
6. Recovery phase

There can be variation in the principle of management of fishery resources, according to phase of fishery. In the predevelopment phase, it is important to promote the fishery. While in the growth or fully exploitation phase, it is important to maintain the fishery. If in case, the fishery is in the overexploitation phase, there should further attempt in order to recover the fishery.

1.2.1. Promotion of underdeveloped fishery

There are a number of different constituents in the fisheries which are usually in various stages of development. Indian fisheries are majorly operated in less than 50 M depth which is in the fully exploited phase. There are certain fisheries such as *myctophid* fishes, oceanic tuna, deep-sea prawn, bulls eye, which remain unexploited.

For the stock, there should be proper management which further aims at creating proper fishing opportunities; further decreasing the pressure of most of the exploited fisheries. It is important to have a greater focus on the investment in the form of fishing gear, vessel which will be best suited in order to exploit the identified stock.

1.2.2. Maintenance of developed fishery

There can be different strategy for the maintenance of the developed fishery in comparison to that of promoting an underdeveloped fishery. Prior to taking any management action, it is important to collect the information about the fish stock and the state of fisher folk, which further helps in the process of decision making, as whether to restrict or further promote the fishery.

In such cases related to the restriction proper methods, it is important to identify as well as adapt the restriction. Throughout the stage, the views of the stock in the future are said to be much more important in comparison to its current status. Socioeconomics of the fisher folks are required to be given priority at such a stage.

1.2.3. Rebuilding of depleted fishery

In order to rebuild the depleted fisheries, it is important to know the cause of depletion. It can either be environmental or fishing. If it is fishing, then some corrective steps should be taken such as a fishing ban or mesh size regulation.

There should be substitute employment opportunities which can be opened up for the fisher folk well before imposing a ban.

1.3. WHO IS THE FISHERY MANAGER?

A fish manager is the in-charge of the overseeing populations of the fish within a region or specific body of water which is referred to as a fishery.

It is the job of the manager to maintain a healthy fish population as well as to take measures when the fish stocks are at risk. In the case of the United States, it can be said that there are both freshwater as well as saltwater fish managers. They work with federal, state as well as local governments in order to create a fishing policy.

The populations of fish can be very delicate and so, a number of factors can impact the wellbeing of the fish. It is important for the fish managers to monitor the environment and watch out for the pollution from dangerous chemicals or runoff. If there is increase in the predatory population increases, then other fish populations will likely to decrease. But probably, the biggest factor is overfishing from humans.

It is important for fish managers to have a deep understanding of the fish they oversee.

They are required to know where and when the fish reproduce. They are required to know the life cycle of each and every type of fish. If the fish are susceptible to some specific diseases or chemicals, the managers are required to be aware of that too.

There are some other factors also which the fish manager should keep in mind when they do their job and one of them is the human factor. The decisions that is taken by the fish managers can have some real-life consequences for a large number of people, from fishermen to consumers.

There are fishermen who depend on harvesting some specific fish stocks for their livelihood and these fishermen can find themselves out of a job if in case, a fish manager places some restriction on fishing. It is not always an easy decision for the fish manager to balance the needs and demands of the community with those of the environment.

There are times, when the fish manager has to deal with some delicate situations. Let's take an example, in the United States, there are some tribal communities who depend on fishing as a means of subsistence or as a way in order to preserve cultural traditions.

There are a number of different areas in the United States which have special regulations which in turn protect these communities and their fishing practices. This can happen sometimes at the expense of recreational or commercial fishing. Fish managers are required to have all into consideration while creating policies.

1.4. GOALS AND OBJECTIVES: WHO NEEDS THEM IN A FISHERY?

There are over-riding goal of fisheries management and that is the long-term sustainable use of the fisheries resources. In order to achieve such a goal, it is important to have a proactive approach and should also involve actively in seeking ways in order to optimize the benefits which are derived from the resources available.

Even though, this happens rarely, the fisheries management is still most common in practice as a reactive activity. In this activity, the decisions are made and actions are taken majorly in response to the problems or crises.

Then the resulting crisis decisions normally attempts merely in order to solve the immediate problems without even considering the broader perspective as well as the longer-term objectives.

Such a kind of approach can achieve success in maintaining the dissatisfaction sufficiently low in order to avoid major conflict, but it is extremely not likely to result in the best use of the marine resources which are being exploited by the fishery.

Figure 1.2.*Fisheries management includes regulation, protection, and conservation of fisheries.*

Source: Image by Unsplash.

In proactive fisheries management, the very first step is to decide the optimization of the benefits for each and every fishery. This is something on which the state or the collection of the legitimate interested parties tend to agree on as being optimal benefits.

In general terms, this can be described in the national fisheries policy which must be the starting point in order to determine the specific objectives for each fishery. There are broad goals which are provided in the fishery policy and this can need to be tailored for a specific fishery, but the goals which are focused for each fishery are required to be consistent with the policy.

In general terms, it can be said that the goals in the fisheries management can be further divided into four subsets named as biological, ecological, economic, and social. In this, social includes political as well as cultural goals.

The biological and ecological goals can be much more correctly thought of as constraints while achieving the desired economic as well as social benefits. But considering the simplicity and consistency with the terminology which are much more usually used in the management of fisheries.

There can be examples of goals under each of these categories which include:

- maintaining the target species at or above the levels which are required to ensure their productivity continuously (biological);
- Minimizing the impacts of fishing on the physical environment as well as on non-target (bycatch), associated, and dependent species (ecological);
- Maximizing the net incomes of the fishers who are participating (economic); and
- maximizing the employment opportunities for those which are dependent on the fishery for their livelihoods (social).

It is important to identify such goals while clarifying the way in which the fish resources are to be used in order to benefit society, and they should be agreed upon as well as recorded, both at the policy level as well as for each fishery. If there were no such goals, there is no guidance on the way in which the fishery should be operated.

This further can result in a high probability of *ad hoc* decisions and sub-optimal use of the resources (which further lead to the lost benefits) and moreover, it increases the probability of the serious conflicts as a number of various interest groups jostle for greater shares of the benefits.

Most of the time, it is seen in practice and one of the important causes of failures considering the fisheries management has been identified as the frequent absence of clear as well as precise objectives.

The very first step which is said to be important is to set goals, the goals which are mentioned above have two obvious limitations. First and foremost, they contain some clear conflicts in intention because of the fact that it is impossible to minimize the impacts of the fishery on the ecosystem and at the same time, to maximize net incomes.

In the similar manner, there are management strategies which aim to maximize net incomes and it is very probable that these management strategies will not also maximize the employment opportunities.

There can be compromise in between these goals which has to be achieved before forming any effective management strategy. One more limitation of the goals is that they are too vague to be of much benefit to the manager.

Let's take an example like the impacts of the fishing can only be minimized if there is no fishing done at all, which is not likely to have been the intention of those who stated the goal.

The employment opportunities can be maximized. This means allowing as many fishers as possible in order to participate without considering the fact that whether or not they could make a living from the fishery. Other than this can also mean maximizing the number which could still earn some acceptable incomes, or many other such targets can be there. Having these examples of goals, there is too much left to the discretion of the manager.

Therefore, it is important to refine the goals further and also, to develop the operational objectives for each and every fishery. There can be operational objectives which are very precise and are also, formulated in such a way that they should be achieved in that fishery at the same time. in other words, it can be said that, the trade-offs between the biological, ecological, economic as well as social goals is required to be agreed upon and resolve the conflicts and the contradictions.

Having such kind of operational objectives, it becomes possible for any observer, which includes manager, to establish goals which can even be achieved or not and hence, there arises a question whether or not the management strategy is relevant and further being implemented successfully or not.

Such kind of operational objectives can also be easily used as the base for the reference points. These are essentially the operational objectives which are expressed in such a way that it can be estimated or simulated in the fisheries assessment.

Once the operational objectives have been agreed upon, there can be development of a management strategy, which is made up of a suite of a number of various management measures, in order to achieve those objectives.

All of this, can sometimes sound like complex. But the reality is different, it is no more than what a lot of people do so as to develop a budget considering their personal finances. A large number of people have realistic but imprecisely expressed hopes and needs or demands for having a lifestyle and also, a knowledge of the nature of the resource (considering the case of net income).Such kind of hopes and needs are the goals of the budget but they all will tend to compete for the same resource that is the net income. So, there will possibly be conflicts which are required to be resolved.

And therefore, it becomes important to modify the goals as well as express them in a much more precise manner. for that, operational objectives are developed which specify what we can realistically achieve considering food, housing, and education etc.

- After this, it is important to decide a strategy on budget which can include:
- How can we meet such objectives?
- What should be the quality and quantity of food and clothing that should be bought?
- what kind of housing should be considered?
- Is it important to consider an annual holiday? etc.

Clearly, it can be said that our operational objectives are required to be consistent with the yield that can be expected from the resource that is the income. Normally, trade-offs are required in the process of developing realistic objectives.

Most of the peoples find that allocation of payments cannot be done much for entertainment or holidays as the people would like and simultaneously, make the rental or mortgage payments.

Then, the priorities are established and further compromises are made until they arrive at some realistic objectives.

These objectives can even balance their desires with their income and furthermore, they provide a good guide on the way in which the finances can be managed from month to month and in the longer-term.

Ultimately, it is important to have a feasible financial management strategy which will have a predictable outcome while barring entirely unexpected events.

If all the calculations are done correctly as well as responsibly, the strategy should mean that the people will enjoy a reasonable lifestyle without even compromising and even being sued for bankruptcy. This is said to be a little different from the basic task and the overall hope of the fisheries manager.

1.5. FISHERIES MANAGEMENT CONSIDERATIONS

1.5.1. Biological considerations

There are living populations, communities, aquatic living resources which are able to renew by using the processes of enlargement in size as well as accumulation of individuals and additions to the population or community throughout the process of reproduction.

Considering a fished population, it can be said that total mortality contains the natural mortality as well as fishing mortality. One of the main tasks in the fisheries management is to ensure the fishing mortality which should not go above the amount which the population can withstand, along with the natural mortality, without harming or injuring the sustainability and productivity of the population.

1.5.2. Ecological and environmental considerations

The changes in the biological, chemical, geological or physical components of the ecosystem can have some kind of effects on the resource population and community. Among these changes, there can be some which can be beyond the control of human such as the upwelling processes enriching some of the coastal ecosystems or large-scale temperature anomalies. But still it is important to consider them while managing the resources.There can be some other impacts which can be because of the human actions also such as the destruction of the coastal habitats for the development, or the direct impact that can be caused because of the fishing on the substrate or on other species further impacting the resources.

1.5.3. Social and cultural considerations

In the fishery management, a most important social restraint can be that the human societies as well as behavior are distorted with much difficulty. There can be case when the fishing families or the communities may not be ready to shift into some other occupations or be absent from their normal homes. When there is a lot of capability in s fishery, even in the situation when their excellence of life may be anguish because of the exhausted fish resources.

Figure 1.3. *Representation of coastal habitat.*

Source: Image by Flickr.

The difficulty is not as good as in comparison to the case, when there are no other opportunities outside fisheries in which they can be get paid so as to meet the important necessity of their living.

1.5.4. Economic considerations

There can be fishery in which the accuracy in sustainable economic capability can be there as the only profit that can be extracted and in which most of the favourable conditions can be prevailed. In such fishery, the market forces can be predictable in order to result in the preferred objective of the economic efficiency.

However, in actual condition, such optimum conditions are hardly established and the uncertainties as well as externalities can negatively affect the normal selection of market forces.

1.5.5. Ecosystem considerations

An ecosystem approach to the fisheries management can said to be the one that is geographically specified, adaptive, and also that considers the ecosystem knowledge as well as uncertainties, while referring to the multiple external influences and further also, strives to the balance diverse societal objectives (Sissenwine and Murawski, 2004).

The effects of the fisheries on the ecosystem and effects of the ecosystem state and variability on the fisheries is recognized by referring to the ecosystem approach (Crowder et al., 2008). There are a lots of articles, which have written about what elements can be included in such kind of approach.

This is because the United States along with many other countries, embraced an ecosystem approach to the fisheries management (Pikitch et al., 2004; Murawski, 2007).

There is one act named as Sustainable Fisheries Act (SFA). The section 207 of which provided the initial guidance on inclusion of ecosystem principles inthe management of the fisheries that are in the nation.

The decision related to this guidance was basically based on the perceived realization that the measures provided by the traditional management were not enough sufficient to ensure sustainable fisheries.

The basic premise is that the fished stocks tend to form important components of the complex marine ecosystems which should be well understood so as to be managed effectively. Some of the guidelines recognize that there is existence of a large number of independent relationships among the stocks, their fisheries and the ecosystem in which they live.

Sustainable Fisheries Act (SFA) mandate the formation of the Ecosystems Advisory Panel of the National Marine Fisheries Service. This was entrusted with reviewing the progress that is made in the direction towards the implementation of ecosystem principles in the FMPs that is Fishery Management Plans (FMPs).

The result was the panel report (NOAA, 1999) which further specified the need to consider and minimize the by catch and discard of fish, identify, and take relevant measures in order to protect the required fish habitat and furthermore to determine the effects of the fishing on the environment.

Despite the fact that the consideration of the factors beyond the single-species dynamics is clearly a major objective, the details of how to do this are still a subject of ongoing research, and there are various approaches which are being pursued.

The US fisheries science centers have conducted much work in order to incorporate the ecosystem considerations into fisheries management (Hollowed et al., 2011).

Such kind of approaches include implementing indices of environmental as well as biological conditions into stock assessment projections (Hare et al., 2010), by further using models which are related to the multi-species and food-when in order to assess the effects of harvesting strategies (Link et al., 2011).

They are also searching for new forecasting methods (Deyle et al., 2013). Such kind of models can be used for providing support to the rebuilding efforts.

Considering the context of fish stock rebuilding, changes have been made in the ecosystem which can further alter the rebuilding rates and the target biomass level till which an overfished species are required to be rebuilt.

1.5.6. Considerations imposed by other parties

There are some offshore fisheries which tend to work in effective separation done from some other users and the local fisheries organizations, which are charged with their management, may be able to run the fisheries in which it is not required to consider the conflicts with or intrusion from the non-fishery users.However, a large number of global fishery landings get nearer from coastal waters and for several others, which are not nearer, tend to produce these landings, these users are a significant consideration and also, a constraint regularly.

1.6. DESIGNING A MANAGEMENT PLAN

What should a management plan contain?

There are certain institutional arrangements which are relevant to a fishery and so, are required while defining the contents of its management plan.

The fishery takes place in international, national, and even regional context and these contexts will influence the fishery policies as well as legislation which govern the fishery.

In some other countries such as the USA or Australia, there are some particular references which are important in order to obtain the management plans in their fisheries legislation (there can be some examples such as the Fisheries Act in Queensland, Australia; Magnuson-Stevens Act in the United States).

That same legislation may specify the main sections which should all be contained in a management plan. At a minimum, FMPs are required to contain:

- A description of the fishery, especially it should include its current status as well as user rights if any established;
- The objectives for the management;
- How these objectives are to be achieved;
- How the review of plan should be done and/or appealed; and
- The consultation process for review as well as appeal.

Requirements in international fisheries policy

In the current scenario, one of the most basic requirements which are recognized in FMPs is the adherence to the United Nations (UN) Convention (which is international sanctioned) on the Law of the Sea (LOS).

On 10th December in the year 1982 (LOS Convention), the FAO Code of Conduct for Responsible Fisheries and the precautionary approach in the direction towards management.

Such sort of adherence should be recognized in the policies related to fishery of each State or in the statutes of international organizations which are in charge of coordinating fisheries management. For the fisheries in the high seas and for those plans which tends to deal with the straddling stocks, it is required for FMPs to clearly specify the international regulatory as well as institutional context in which the plan is applicable (an example can be of UN fish stocks agreement).

Considering such cases, FMP is required to have clear links to both the national fishery management policy as well as to the statutes of the commission or international organization which are in charge of coordinating the management of the international fishery.

Let's take an example. The Australian Fishery Management Plan (FMP) for the southern bluefin tuna contains its first objective as: 'in order to ensure the conservation of the stock of southern bluefin tuna, it is important to have an appropriate management and also, along with the Commission for the Conservation of Southern Bluefin Tuna.'

Figure 1.4.*Image of bluefin tuna.*

Source: Image by Flickr.

In reference to the national FMPs, there is a word of caution, that should be expressed, which is designed in order to manage the fisheries for the resources which are distributed in the high seas or which are shared in between two or even more countries.The FMP is a national instrument and therefore, these resources are not under the jurisdiction of a single nation. The main reason for the same is that the FMP will only be applicable within the EEZ of that nation.

This not at all means that the FMP will not be effective but it means that the plan can be required to define national objectives which are constrained by the objectives of some other countries.

1.6.1. National/State fisheries policy requirements

It is important for the National fishery policies to get support from a legal as well as institutional framework, so that FMPs can be considered as the main instrument of management. With reference to this, every national fishery policy is required to define the range of minimum requirements which should be fulfilled by the fishery management plan.

Therefore, the national policy is required to broadly determine the type of information which should be included in an FMP so that all the FMPs which are obtained can achieve the goals of the national fishery policy.

Since, these goals or aims are much likely to be broad, the information requirements are much likely to be general. But despite this, it remains the responsibility of those who are preparing the FMP to ensure that the particular objectives of an FMP should comply with the national fisheries policy.

1.6.2. Fishery-specific requirements

There are some minimum requirements which are specified by national fishery policies and these requirements tend to still leave the significant leeway in the contents of an FMP. This leads to large difference in between the FMPs which are developed for different fisheries even if it is within a single fisheries jurisdiction.

It is important to consider this as a strength of the policy. The reason being it ensures that the FMP is developed in order to suit the particular needs or requirements of the managers of each fishery and not just become a bureaucratic requirement.

It is important that the plan should specify the way in which the management objectives (that is operational objectives) of the plan are to be achieved. Each management objective should be examined individually if in case it is possible and the identification and explanation of such objectives should be done which can be achieved by designing the management measures.Here, it is important to keep the needs of the plan very specific on the way in which the objectives and measures link up and which the performance indicators are going to be used in order to measure the achievement of the objectives of management.

It is important to do this and this can be done by structuring the plan in such a way that it requires the development of performance indicators.

The indicators are required to cover all the objectives of the plan. So, generally, there will be indicators of the biological status of the stock but also, there will be social as well as economic indicators of the health of the fishery.

Such kind of indicators will equally reflect the agreed reference points and also, it can measure how successfully the target reference points are being achieved by the plan and remaining within the limit reference points. In general terms, it can be said that the plan will not specify the exact way in which the indicators are developed, but the development of these is also important.

Let's take an example. The forecasts on the economic impact of the actions taken by the fisheries management can be used as an indicator of success of a plan while trying to achieve the objectives that is of the maximization of economic returns.

The FMP is required to specifically make reference to the characteristics of the natural environment in which the fishery take place and to the way in which the changes in such an environment can affect the management of the fishery.

If there are traditional management structures or some established rights which have been used in previous times, these are required to be recognized and furthermore, included in the FMP.

There are management objectives in an FMP which conflict most of the times such as resource conservation vs maximizing economic returns or promotion of the development. It is important for the plan to acknowledge this conflict and address the same through requiring a process by which conflicts in between the objectives can get resolved.

This can be further achieved by specifying some of the indicators which are related to the conservation objectives as constraints, whereas the identification of the indicators of the economic performance can be done as targets.

This can lead to the development of operational objectives and can result in appropriate target and can further limit the reference points. Considering this context, it can be said that the development refers to the human resources, and therefore, it encompasses capacity building and further can enhance the quality of life.

1.7. FISHERIES MANAGEMENT TECHNIQUES

It is important to manage the fisheries resources in order to make sure that the stocks are harvested at sustainable levels for the benefit of present as well as future generations. For example, the fisheries resources which are located in the New South Wales are very diverse.

There are a large number of species which are highly regarded by recreational fishers for sport as well as food, by the commercial sector for their monetary value and by the community as fresh, quality seafood.

Figure 1.5. *Different types of seafood.*

Source: Image by Unsplash.

NSW Fisheries are considered as custodian of the resource and therefore, has the responsibility to manage the fish stocks on behalf of the community.

There has been a perception which held from a long period because of the fact that the commercial fishing sector catch a large number of fishes and so, it is important to be heavily regulated. Whereas, this can be the case in some of the circumstances, there is recognition which is increasing.

They also, that an estimated 1 million recreational fishers, who fish in NSW in each particular year, have a considerable effect on the fisheries resources. Therefore, it seems to be important to manage the recreational sector.

1. Commercial fisheries management

Throughout Australia, the fisheries agencies have traditionally categorized fishery resources into some segments which is basically for the ease of management. This is specifically important in NSW and the reason being that the large number of species are taken for sale and the range of gear types are there which are used by the commercial sector.

Historically, the commercial fisheries have been defined by using the method of capture. Let's take an example that is prawn trawl and fish trap. However, there are some of the more valuable fisheries having one target species which are further defined by the species itself such as abalone and rock lobster.

Once each and every fishery is defined, there can be policies and legislation which can be developed in order establish some particular rules. Such kind of rules can then be included in a management plan for each and every fishery. A fishery management plan can be developed and developing the plan can include the extensive consultation with the commercial sector, the recreational sector, and the community.

Considering most of the cases, a specialized committee is established in order to develop a management plan which is acceptable to all. This committee contain NSW fisheries management as well as research staff, commercial fishers and recreational fishers.

2. Commercial management tools

Broadly, there are two types of fishery management tools such as input controls and output controls. Input controls are said to limit the amount of effort that are done by the commercial fishers into their fishing activities, by controlling the amount of fish caught indirectly.

Input controls can include restrictions on the number of licenses, the size as well as engine capacity of boats, the length as well as mesh size of nets and the areas and times which can be worked. Considering some locations, there are commercial fishers which are not permitted to work during the weekends or public holidays.

There can be a good example of how input controls work and that is the juvenile king prawn closure near the mouth of the Richmond River at Ballina. King prawns live in the estuaries like the juveniles do and further they tend to migrate to the ocean during the young adult stage of their life cycle. This makes them vulnerable so as to get captured at a smaller size by prawn trawlers.

In order to protect the stocks of smaller prawns and also, for increasing the value of the prawns sold, the prawn boats are prohibited and they are no more allowed to trawl within an area bounded by a 2.2 nautical mile radius from the mouth of the Richmond River. Such kind of management measure also helps in reducing the capture of juvenile fish in prawn trawl nets.

On the other hand, the output controls limit the amount of fish directly which can be taken from the water. While implementing an output control management regime, the first step is setting a Total Allowable Catch (TAC) for the species.

When this is once established, the TAC that is Total Allowable Catch can be either fished on a competitive basis or divided up between the participants in the fishery. This is done so that all fishers have an individual quota. A TAC is divided between the participants which in turn means that in any one year a commercial fisher is not be able to catch more than their allocated weight of that species.

Generally, the output controls are regarded as good mechanisms for controlling the total catch in single species, high value fisheries such as abalone and lobster, which are being targeted by using a single gear type. However, there are thorough monitoring schemes such as daily catch logbooks which are required most of the times in order to ensure that individual quotas are not exceeded.

3. Recreational fisheries management

There are a lot of goals of recreational fisheries management which do differ from those of commercial fisheries. The reason is that they have the broader aim of providing the optimum use and recreational enjoyment of fisheries resources.

The management arrangements are brought into being both for controlling the total harvest, and also, for sharing the angling experience amongst fishers.

There are a number of restrictions on the types of recreational fishing gear which are allowed. The major restriction is the anglers and that is to line fishing methods having limited hook numbers, and a few small nets.

Such sort of restrictions operates as input controls, as do closures on fishing in some of the certain areas (such as marine reserves) and closed seasons (such as Murray cod spawning season).

The output controls for recreational fisheries can take the form of daily allowable catches (that is referred to as bag limits) and size limits. There are bag limits which are set in order to limit the total harvest of some specific species to sustainable levels, and also, to share the catch.

When there is absence of comprehensive scientific data, then they usually depend upon an amount of fish which is considered as socially acceptable that is to be taken in one day (such as enough to feed the average family). Let's take an example and that is the anglers are only allowed to keep ten snapper per day.

A minimum size limit is set and that is a management tool. This management tool is regularly used in order to ensure that most of the fish in the population have the ability to grow to breeding size and spawn at least once before they are caught. An example that can be considered is that, the size limit on turban shells in NSW is 7.5 cm, that is a length at which 100% of the shells sampled were of reproductive age.

These are the attempts which are made in order to make sure that there will always be availability of juvenile fish in the years to grow and further enter the adult population. There is a maximum size limits that applies to certain species because of the fact that the larger fish breed most of the time in a more efficient manner in comparison to smaller fish (e.g., the anglers may only keep two mellowest above 70 cm in length, per day).

The NSW DPI determines the size limits for each and every species while considering the scientific advice about the biology of fish stocks, and also, considering the consumer preferences and the existing sizes that is taken by commercial and recreational fishers.

4. Habitat management

For long term, the fisheries management will only be effective if the fish habitat is protected. The coastal population is always increasing and this has resulted in increased land and water pollution and further having greater pressure for the coastal development.

Such an increased pressure affects the areas which is near the edges of estuaries and bays. Both of which are important in order to survive for juvenile fishes and crustaceans. Considering the Fisheries Management

Act 1994, NSW Fisheries plays a larger role in deciding whether various developments can proceed or not.

5. Education and enforcement

There can be decrease in the effectiveness of the fisheries management if there is no adequate education and enforcement of the rules. Commercial fishers, anglers, and the community are required to understand the reasons for and obeying the fisheries policies, legislation, and management plans in order to make sure the resource is exploited at safe as well as sustainable levels.

1.8. FISHERIES MANAGEMENT SYSTEMS

A fisheries management system can be defined as the institutional structure as well as administrative routines which intends:

- To generate the information which is required to make effective decisions considering the use of limited financial and personnel resources having the purpose of optimizing benefits from fisheries resources for society;
- To make such kind of decisions, and
- To implement them and furthermore, assess their efficacy.

There are three basic levels of conceptualization of fishery systems which are used in fisheries management. At the very first level, a fish stock (that is the resource) is conceptualized by the prevailing fisheries science paradigm.

This fish stock is affected by fishing activity (that is the fishing effort). The focus of the management is on the harvesting sub-sector and, more specifically, on each of the particular fish stocks which are being targeted.

The focus of management is on fishing effort that arises from an assessment of state of the stocks. Furthermore, the information is gathered by management on the fish stocks as well as fishing effort.

Then, this information is used as the basis for decision-making. The focus of management is created as controlling fishing effort in order to achieve a particular catch as well as stock level.

There is a range of biological (Schaefer, 1954), bio-economic (Hannesson, 1992) and bio-socio-economic (Panayotou, 1982) models which are further related with this paradigm. These share a number of common assumptions as well as features.

In particular, the focus of management is the resource and the issues (problems along with solutions) and these are considered at the level of each and every fishery. A number of main assumptions are derived from the 'tragedy of the commons' paradigm.

At a second level, the human sciences approach to the conceptualization of a fishery system and also, this has three basic components in the harvesting sub-sector and these include the resource (that is the fish stock), activity (that is the fishing effort) and the actors (that is the fishers).

Such kind of approach mainly focuses on the management of fisheries and that is primarily a problem about the behaviour of people and not about the behaviour of fish.The actor (that is fisher) is considered to be the prime focus of management, and there have been a number of different attempts which have been made to explore, explain as well as predict actor behavior through economics (Cunningham et al., 1985), institutional economics (Wilson and Lent, 1994), social, and political economy (Jentoftet al., 1998) and cultural economics (Acheson, 1981).Considering a third level, the fisheries systems approach sets the harvesting sub-system (that is the actor-activity-actor) and the management authority in a much larger total system. This contains two major spheres namely the physical bio-sphere and the social sphere. There is a much wider system of interactions in the bio-physical sphere which is embedded with the behavior of fish stocks.

There are three issues of importance which do emerge from this exposition of the different approaches to the conceptualization of fisheries systems and therefore, for the fishery management systems that is based on them.First of all, there have been progress in the conceptualization of fisheries systems which have been made from a simple fisheries science-based approach to the more sophisticated and all-embracing systems approach. The second one is the limitations of the fisheries science approach in order to provide adequate fisheries management advice and so, are now better understood.

The third point is the possibilities in order to develop appropriate fisheries management systems on the basis of multi-disciplinary approaches. This, in turn, reflect the complexity as well as dynamics of the system, and so, this have been widely recognized.

1.8.1. Implications for knowledge required and institutions

Fishery systems are said to be complex, and there is a need of management systems which is required to optimize the benefits that is accrued from fisheries. This require institutions as well as knowledge systems which are able to cope up with the multidisciplinary requirements of the fisheries management function.There is an indisputable need of building the capacity of fisheries institutions in order to identify, generate, deliver, and utilize knowledge which is appropriate to the fisheries management. However, the level of complexity of the knowledge is daunting, and this is the knowledge that is required whereas some are very expensive to obtain.

This immediately puts up the question and that is how the needs of this knowledge can be met under conditions, the condition where a number of fisheries management authorities do not have the financial as well as personnel resources in order to meet a more limited spectrum of knowledge requirements.The foundation should be the best available knowledge upon which the decision-making in policy formulation and monitoring is done. There should be knowledge within the realm of what is realistic and what is attainable.There is a useful aspect of the systems approach and that is to understand the working of a fishery. It encourages and performs broader as well as more realistic analysis of the fishery. However, there should be decisions which should not be delayed pending the availability of knowledge that goes well beyond the availability of the capacity of the expertise and finances.There is a need for fisheries managers to make decisions on the basis of imperfect, even though the best knowledge that is available is closely related with the application of the precautionary principle.

1.9. CONCLUSION

Fisheries management can be referred to as taking care of fishery wealth so that it is possible to use them sustainability, drawing on fisheries science and furthermore, counting the precautionary principle. The fisheries management authority is said to be the unit which are allowed by the team to perform exact management functions.

There are a multifaceted and wide-ranging set of tasks which is included in the fisheries management, which jointly have the attainment of optimal profit that is sustainable from the resources as the underlying goal.

There are fishery managers who tend to take important decisions regarding the fisheries or even its management. Also, there are fisheries management systems which is referred to as the institutional structures which includes generation of information, making the decision and the implementation of the information.

REFERENCES

1. Cochrane, K., (n.d.). *Chapter 1: Fisheries Management.* [online] Fao.org. Available at: http://www.fao.org/3/y3427e/y3427e03.htm (accessed on 8 March 2020).
2. David, D., (n.d.). *Chapter 9: Design and Implementation of Management Plans.* [online] Fao.org. Available at: http://www.fao.org/3/y3427e/y3427e0b.htm (accessed on 8 March 2020).
3. Dpi.nsw.gov.au. (n.d.). *Fisheries Management Techniques.* [online] Available at: https://www.dpi.nsw.gov.au/fishing/commercial/info/fisheries-management-techniques (accessed on 8 March 2020).
4. Key concepts I: Fisheries management systems and governance, (2004). [eBook] p. 1. Available at: http://siteresources.worldbank.org/INTARD/Resources/335807–1323958783967/8321851–1323959552668/WBPolicyBrief12.PDF (accessed on 8 March 2020).
5. Lavanya, B., (2017). *Fisheries Management.* [eBook] p. 2. Available at: https://symbiosisonlinepublishing.com/marine-biology-research/marine-biology-research17.pdf (accessed on 8 March 2020).
6. Ramteke, K., Joshi, H., Bhendarkar, M., Renjith, R., Ingole, N., & Hapse, V., (n.d.). *Basic Principles of Fishery Resource Management.* [online] Aquafind.com. Available at: http://aquafind.com/articles/Principles-Of-Fishery-Management.php (accessed on 8 March 2020).
7. Rowe, W., (2019). *Fisheries Management | List of High Impact Articles | PPts | Journals | Videos.* [online] Imedpub.com. Available at: http://www.imedpub.com/scholarly/fisheries-management-journals-articles-ppts-list.php (accessed on 8 March 2020).
8. Strickland, J., (n.d.). *What are Fish Managers?* [online] HowStuffWorks. Available at: https://adventure.howstuffworks.com/outdoor-activities/fishing/fish-conservation/fish-populations/fish-manager.htm (accessed on 8 March 2020).
9. The National Academies Press, (n.d.). Ecosystem Considerations. [online] Available at: https://www.nap.edu/read/18488/chapter/7#86 (accessed on 8 March 2020).

2

Property Rights in Fisheries and its Management

CONTENTS

This chapter is about the rights of fishery and the way to manage these rights in the society. This chapter explains different rights that are given to the public or fisherman that gives them the right to have fishing in the particular area but with some rules and regulations.

These rights are being developed to make the easy availability of source and to avoid the wastage. These rules or rights help the government to maintain the importance of fishing by allowing only particular people to have the catch.

This chapter tells the importance of maintaining these rights and how their implication process takes place. It also tells the importance and effect of limited licensing. Also focuses on the rights that are made for the fishermen and their importance in the society with their management.

Figure 2.1. *Fishermen preparing themselves for fishing.*

Source: Image by Dreamstime.com.

2.1. INTRODUCTION

Nowadays, it is being recognized that there are lots of problems in the economic inefficiency in the field of fishery (Scott, 1955; Neher et al., 1989; Shotton, 2000). It is being referred as a most common property problem (Gordon, 1954).

It can be more dramatically said as the tragedy of the commons (Hardin, 1968). This happens due to lack of property rights that might lie under natural resources. These rights basically included fish stocks and the ocean habitats.

In this chapter, basic rights are being discussed with the management process. There are some of the issues that need to be consider:

- Determination of total allowable catch;
- Monitoring and enforcement;
- Nature of property rights;
- Need for other regulations;
- Management units; and
- Rent extraction, cost recovery, and initial allocation.

When there are restrictions on the ownership and transferability in ITQs-based management, there is normally increase in the programmed complexity and it also reduces flexibility for the individual participants.

When these restrictions are considered, then it is important to ensure that the trade off in terms of other fishery management objectives is worth enough to the cost of the increased complexity and the reduction in the gain from the restricted flexibility.

Now, everyone prefers fishing as "Right based fishing." The main reason using this as a recent fishing method because of the failure of the traditional fishing management. This method is done to improve the economics of fishing industries.

Figure 2.2. Determining specific amount of catch quantity.

Source: Image by Pxhere.

Traditionally when there was excessive pressure on fish stocks, they usually prefer to reduce the use and efficiency of fishing vessel, this was done to protect the resources. But this technique directly affected economic point of view. The availability of overcapacity of fishing fleets is also considered as a constant threat to the health of fish stocks.

2.2. IMPORTANCE OF PROPERTY IN FISHERIES

2.2.1. What does property in ocean means?

Hugo Grotius gave a good common-sense explanation. Hugo was an international lawyer and philosopher. Hugo explained that two conditions are there for holding property in a thing. Firstly, to have power to appropriate the things and to have hold in possession against others. Secondly, the things had to be scarce and exhaustible.

What powers has the holder of a property right?

There are three rights that can be found in any of the system of law regarding land or water:

1. Power to use things (manage it)
2. power to dispose it (sell it or grant it)
3. Power to take its yields (as a crop, rent or royalty)

2.3. PROPERTY RIGHTS IN THE FISHERIES SECTOR

In the current fishery sector, the concepts of property rights need more specific definition. These rights are helpful to regard them towards concerning a resource flow in terms of harvest and catch over time. In the year 1991, Bromley suggested that "Property is not an object such as land but is rather a right or group of rights to a benefit stream that is only as secure as the duty of all others to respect the conditions that protect that stream."

Figure 2.3. *Fishermen fishing in their specified located areas.*

Source: Image by Wikimedia Commons.

For natural resources, such approaches are applied in case of all the management regimes, but not just in case fisheries. When an open access is considered, which provides a way to understand property rights in a better way.

Here open access means that everybody can exploit the natural resources in the fish stock in free. In this situation, no one excluded from the pool, this type of situation is often called as 'common property' which is basically misunderstanding the concept of Bromley, 1991. Whereas a common property stands, where certain group is only allowed to take the benefits which excludes the outside groups.

Open access is applicable only on those places where there are no rights recognized in an exhaustible resource.

There are basically two conditions that are needed for the existence of property rights:

- Firstly, exclusion of those who are outside the owners, groups or users and to have protection of those rights by using authorities. The absence of protection and enforcement can lead to a situation of open access in practice.

This gives a definition 'A right is the capacity to call upon the collective to stand behind one's claim to a benefit stream (that is to one's property), (Bromley, 1991).'

Bromley (1991) indicates four broad regimes that are reflecting ownership of property. The constraints implicit in the associated rights. These are in altered form:

1. **State Property:** in state property managing agencies have the rights that can determine the rules of access or used, and it becomes duty for an individual to observe them.
2. **Private Property:** in private property, individuals are having rights for those who are socially accepted users of their property and duties. This is done in order to avoid the socially accepted uses and the rights to expect only those users.
3. **Common Property:** common property is where in addition to the relevant elements those are relating to the private property. It is being socially accepted uses.

It is important to maintain the balance between the groups and rest of the communities as well and within groups it is maintained in terms of using and maintaining the resources.

4. **Non-Property:** non-property is where the individual is supposed to have no rights but the resources are open to all subjects. This is done only to constraints of a wider nature that relates to individual behavior in a society.

Figure 2.4. Different constraints with their associated rights.

2.4. CHARACTERISTICS OF PROPERTY RIGHTS

Scott (1989) suggested that the bundles of rights with the associated rules, are being evaluated in terms of the presence and the absence of certain characteristics. Scott listed six presumable arbitraries which includes duration, flexibility, exclusivity, transferability, diversity, and quality of title. Quality of title focusing on the dominancy which questions about who really owns the property, which is consider as critical.

Quality is meant best only where ownership cannot be disputed and is worst in case where there is uncertainty and conflicts. All the other dimensions can be self-explained, except the flexibility. Flexibility which doubtlessly means that the property which is constrained as how it is used.

Therefore, according to the Scott in (1996, 2000) the most crucial property rights are:

- Security or title quality;
- Exclusivity;
- Transferability; and
- Permanence.

2.4.1. Security or Quality of Title

The property rights are included in this type of security rights that are mainly challenged by other individuals and by government or by institutes.

Security term refers to the ability by which owners are said to withstand with these challenges. Together with this they are also said to maintain their property rights. As the owners will have hold on their property, so it is considered as the best thought.

This probability can range from zero to one. When the security measures range with one, then that means that the owner will hold their property with complete certainty. Security measures with range of zero, means that the owner will lose their property.

2.4.2. Exclusivity

Exclusivity refers to the ability by which the property holder utilizes and manages the resources in their own property without letting outsiders to interfere.

An ITQ holder have a right that specifies the volume of harvest that is from the given fish stock over a certain time period. Therefore, ITQ rights generally provide less than 100% of the exclusivity towards the relevant assets.

2.4.3. Permanence

Permanence is the time span of the property rights. Permanence rages from zero. In this case, for infinite duration the property right is values nothing. Example for the property rights of a finite duration is leases.

It should be noted that there are important differences between the indefinite duration. Where property right is perpetuity. This certainly specifies that the property rights will last forever. Thus, for instance, a rental agreement may provide a perfectly secure property right for a limited duration.

2.4.4. Transferability

Transferability is referred to as the process where property rights are transferred to someone else. This is economically important because it facilitates the optimal allocation of the resource. That can compete the users as well the uses.

Divisibility is an important feature of transferability that is the ability to subdivide the property rights into smaller parts. This is done for the transfer purpose. Perfect transferability can be implied by both, no restrictions on transfers and perfect divisibility.

2.4.5. Quality of property rights

Scott (1988) followed the Q-measures. It is helpful in visualizing the characteristics of the property rights. It can be measured along the axis in four-dimensional spaces. It is very convenient to measure the scale which ranges from 0 to 1.When the measure range zero, then the property rights are said to hold none of the characteristics. The term 'quality of property right' refers to the ration which is always positive or less or qual to unity. The higher quality can be achieved by the closer characteristic footprint of a property right to the perfect property right.

It is useful to have a unidimensional numerical measure of the quality of a property right which is given the multi-dimensional nature of the property rights. Such measures are called as Q-measures for the property rights (Arnason, 2000). The below expression can define the Q-measures:

$$Q \equiv S^{\alpha} \cdot E^{\beta} \cdot P^{\gamma} \cdot (w_1 + w_2 \cdot T^{\delta}),$$
$$\alpha, \beta, \gamma, \delta, w_1, w_2 > 0 \quad \text{and} \quad w_1 + w_2 = 1$$

In the above expression, α; β; and δ are parameters and w_1 and w_2 are weights, where the Q-measure are said to take the values in the interval ranges [0, 1]. A value of zero means, the property right has no quality. A value of unity means, the property right is perfect.

2.5. THE NATURE OF THE PROPERTY RIGHT

For the rights-based management programs the most basic attributes are how the property right is denominated. There are basically three types of denominations that are area, input, and outputs. Rights can be arranged to utilize a designated geographical area.

A considerable discussion on the form of property rights is given by Francis Christy which is called TRUFs that is territorial use rights in fisheries (TRUFs). This normally includes the control of the basic decision which focuses on how much to harvest and how the harvesting can be accomplished.

Rights can be denominated in terms of input. In this case the owner is granted with the rights that uses certain inputs in the designated areas with specified times. Generally, this form will be a license program where the number participants are limited.

This is more frequently, are the rights-based on physical harvesting capacity and is measured in terms of attributes of boats such as units of displacement or horsepower, traps, or days at sea, number of boats, etc.

Finally, the rights can be denominated in terms of outputs. In this, the owner is granted with the rights to harvest fish each year with a specific amount. These types of programs are called as individual transferable quotas (ITQs).

Basically, in this case, the total allowable catch (TAC) is divided into small parts and then it is allocated to individual. For the individual owner's flexibility these shares are bought sold and leased too. This helps the owners in planning their fishing activities.

Three words are individual, transferable, and quota. These terms comprise the term ITQ which is important for describing the different orders.

2.5.1. Quota

ITQ programs are basically based on an annual Total Allowable Cash (TAC). The biological concerns for the size of the fish stock and changes over time are addressed by the limiting the total amount of annual fishing mortality.

These procedures help in setting the annual TAC in an ITQ program. Which is fundamentally same as for any other TAC which is based on the management regime.

2.5.2. Individual

For an open race for the TAC, an ITQ program is allocated shared to individual participants. Where each participant has given a right to harvest certain amount of fishes every year. This type of program is useful in distinguishing the share rights itself, which is permanent.

The Annual Harvesting Rights (AHRs) allows a specific amount of harvest each year. To have basic rational for the individual quotas, it is necessary to ensure that the TAC is maintained and flexibility is maximized and individual controls the participants at the same time.

Most of the traditional management programs allows the interested parties to participate. But when they need to control the total harvest, then they operate or shut down the fishery. In the basic premise of ITQs is to regulate the amount of fish each participant can take. But it also allows them to catch fish by specifying where, when, and how.

2.5.3. Transferable

In order to increase the flexibility for the participants, individual quotas share is transferred. In this individual can buy, sell or lease the rights to catch the fish. This gives freedom to the participants to operate at the scale which is more advantageous.

2.6. ISSUES RELATING TO THE NATURE OF THE PROPERTY RIGHT

2.6.1. Eligibility to own

Preventing corporations or general public from participating will help in maintaining industry and community structure. It can also prevent the possibility of the absentee. At the same time, limited flexibility will prohibit right owners from having the opportunity to organize their activities and with that effect the efficient use of fish stocks. With this to accomplish objectives of other fisheries management.

i. allow any legal entity the right to own rights;
ii. Allow only persons the right to own its shares; and
iii. establish other restrictions on eligibility, examples individuals from certain areas or who use certain types of gear can own rights or banning foreign ownership or requiring that only bona fide fishermen.

2.6.2. Duration of ownership right

Here the question arises, should the ownership right be permanent? Or Should it be for some limited period? In some of the countries like the USA, programs like ITQs provides ownership right which guaranteed only as long as the management plan.

This implements the programs in effect. If the fishery management council develops a new plan that meets all the standards in the law then, ITQ program can be eliminated. This includes these changes will be an improvement for the status quo.

But in other countries, ITQ programs are the part of the fisheries law. This allows the owner's right to be as permanent as the current policies allows the right owners that will have the securest possible planning horizon.

This will also have better incentives that will make efficient investments in harvesting and processing equipment in order to develop market channels.

i. ownership rights can be assigned permanently;
ii. Ownership rights can be temporary initially but after a trial period made permanent;
iii. Ownership rights can be assigned for the duration of the existing management plan; and
iv. Ownership rights can be assigned for fixed periods subject to renewal if specified criteria are met.

2.6.3. Transferability I

Fundamental issues arise which questions that whether the rights should be transferred. Some of the people says that making the rights transferable is not appropriate.

They think so because it is a permanent and said to be as a free consignment for a private individual of a public resource. In these terms, transferability offers the possibility of certain individual to obtain wealth from a public resource.

Therefore, these restrictions on transferability would obligate the flexibility of the owners. this is one of the potential advantages of rights especially ITQs.

i. In the case where rights can be leased but not sold, in ITQs the share-right may not be sold but the AHR annual harvest right is transferable perhaps with some restrictions. This will allow flexibility for emergencies such as illness or vessel breakdowns;
ii. In case where the rights can be made non-transferable, the individual who receives the right is the only one that may harvest the fish;
iii. In the case of ITQs, both the share-right itself and the AHR can be fully transferable by sale, inheritance, lease, gift, or through joint-harvesting arrangements; and
iv. In case where rights are sold not leased, then in ITQs the share-right may be transferable but the AHR may not be sold. This will prevent the emergence of absentee.

2.6.4. Transferability II

When the rights are need to be transferred, then some of them will argue that the limitations on the type of trades that can be permitted to justify in certain circumstances.

Their initial allocation includes individuals who normally differs by some factors like, gear type, boat size, firm size, type of final product, homeport, etc.

There are some changes in the industrial and cultural aspects of fisheries that results in the free transferability among all such individuals.

i. Place no restrictions on the transferability of rights among different groups of owners; and
ii. Determine critical groupings of participants and prohibit or restrict purchases leases etc. between members of these groups.

2.6.5. Ownership caps and restrictions

When the ITQ programs are implemented there is possibility that certain entities are obtained as excessive shares. The problem can be relevant for other type of rights as well.

It can be normally referred as a market power for vessel fish, final products and even for ITQ shares. It can be generally referred to historical industrial and cultural make-up of the fisheries.

i. Leaving excessive share problems to antitrust law;
ii. Placement of caps on the percentage of ITQs for a particular management plan. Also, for a particular stock within a management plan that any one entity can own; and
iii. Place other restrictions on how firms or individuals can combine to harvest the ITQ.

2.7. IMPLICATIONS OF ITS MANAGEMENT FOR THE FUNCTIONS

2.7.1. Strategic policy planning

Strategic policy planning involves the responsibility for managing the marine resource. Strategy policies lies between the rights holding harvesters, other stakeholders, political process. This may approve the management regime and the infrastructure that delivers other management function.

Strategic policy planning must articulate the objectives. That is for the management and secure agreement on those objectives which are from all the stakeholders with an expressed interest.

The development of legal base is needed to pursue the objectives. It should ensure that the management infrastructure is funded from general taxes, directed taxes or cost recovery.

The political, socio economic and resources are need to be characterized to determine the management complexities. In unitary states, these are most straightforward situations that are found that helps in managing the resource within the EEZ. In these countries, a single level of government has clear and undiminished jurisdiction and control.

When the states get involved in the management of the stocks when shared with the other management regimes increases administrative and management complexities.

Example, straddling stocks. When federation with national and provincial levels of government shares jurisdiction over resources.

There are other areas of laws that are affecting resource management which includes taxation, safety, trade, and many more. This results in a complex counselling and decision-making processes with longer time frames.

Every legal framework provides the authority to implement policy. Authorities are therefore required to create property rights. These property rights are to recover the cost or collect rents and to make regulations or operational policies.

Fish rights are the access rights rather than allocation of physical property. These physical properties are defined quantitatively and are also depend on the information that defines their limits and, they quantify their uses.

The effectiveness of an ITQ management system depends on the integrity of the information system.

There is a requirement for the data that is accurate, complete, and are also made available in a timely manner for management purpose. This type of obligations should be specified in the legislations.

ITQs are typically used for the management of commercial fishing. The locals or native people must have their food claims or customary. Traditional users are said to have longstanding dependencies that may require protection.

The right which describes 'public right to fish,' often takes the form of recreational access which is a feature in many fisheries.It is usually felt difficult to accommodate these users with individual quotas. Larger number of low intensity users can exceed the administrative capacity of ITQ systems. In order to overcome from this, the shares of the TAC can be allocated to the co-operative or to these groups that are used to represent them.When the subtraction of these share from the TAC takes place then they leave the reminder for the ITQ holder as Total Allowable Commercial Catch (TACC).There are decision rules that are required for setting annual Total Allowable Catch limits. These rules should be preserved in policy. This involves the use of a limit reference point. These reference point makes the minimum level permitted specific for the biomass of the stock. Conversation policies can be specified for the target exploitation rates.These exploitation rates vary depending upon the resource status. The total allowable cash limits can also include the re-building measures for the overexploited stocks. It also contains principles that is to be applied in order to protect associated or dependent species.It also contains the principle that is to be applied to protect habitats. There are efforts made by many of the jurisdictions in order to incorporate the use of ecosystem objectives. While setting exploitation limits, they apply the precautionary approach.In order to make the right holders to participate, ITQ systems required a high level of support. Most of the jurisdictions are now using ITQ management.These ITQ management system is applied in order to have a gradual approach to their introduction which begins first in fisheries. They are considered to have a high likelihood that have successful management approach.

Figure 2.5. Fishermen preparing for fishing.

Source: Image by Pixabay.

For the candidates who are just beginners, for them single species fishery with relatively few vessels is good because that help in easy monitoring. This helps in providing a base that is for extending the system. The legal framework and policies should be used in other to meet these approaches. With experiences, it is suggested that imposition of an ITQ system on a fishery, this increases the risk of management failure.

Most of the jurisdictions who are managing fisheries are supposed to have much of the institutional capacity which supports the strategic policies planning that function in a department or ministry. Right based management system introduction requires leadership and the investment of political capital in order to make the main policy choices and in order to support them with the appropriate legislation and regulations.

Depending on the circumstances and complexity of the fishery these are several processes that can take several years. Many of the rights-based systems are currently in place. They were initiated at the time of crisis when there is a need for fundamental changes in the management rose. These changes take place from top of the political agendas and considerations of the administration cost, which is given secondary importance.

When dealing with the stakeholders, a structured approach should be taken to consultation. This way is the best way to sperate the policy and governance issues from the operational issues. The involvement of the broad stakeholders is said to be appropriate in order to govern the issues that can be solves using strategic level. Operational issues are referred to the operational planning function.

It is considered that operational issues can be best solves by the right holders and locals or it can manage by immediate stakeholders. When the issues are separated in this way then this allows meaningful agendas to be set. This way protagonists can help in bringing value and closure to the topics. Well this separation also helps in promoting the efficiency and reduces the transaction costs.

Comprehensive policy framework is considered best in place prior to the implementation of the ITQ system. It is very common to have a launching of an ITQ program with many of the policy issues which may be unsolved.

At the time when a new system is being introduced, then the policy function will require a greater infusion of resources. With this, there are ongoing requirements that are for the function, in order to maintain the relevant strategic perspective and with this to maintain policy framework over time.

2.7.2. Resource research and analysis

For Fisheries management, it is required to have a biological understanding of the targeted populations. With this, they also required to maintain their relationship with the environment. For fisheries management is important to see the impact of fishing upon them. These sets of information's are especially critical where the level of exploitation is high.

There are many countries those are maintaining an appropriate institutional capacity in the field of marine science and they invest in order to acquire the information that is needed for research. Some of the counties uses input control system that does not require the setting of catch limits. But they are needed for the remain information.

At the time of research, the focus was typically on understanding the biological and environmental process. They also focus on the result while applying input controls. For management and analytical stock assessments the global or Olympic quotas are used that are required to produce TACs.

An ITQ system make the analytical assessments essential. With this they also increase the demand for the assessments that require TACs that to be set for a new fishery. They can also establish fisheries by updating more frequently.

Stock assessment is a part of marine biology which mainly focuses on the dynamics of exploited populations. Stock assessment includes study of biological and population processes which includes growth, recruitment, reproduction, and the mortality. Stock assessment is used for describing the past and present status of fish stocks.

It also forecast the future trajectories of stocks which is under different exploitation scenario. As stock assessment focuses on direct applications for fisheries resource conservation, with that in addition to this they also focus on the population process and to provide information as a key component of the marine ecosystem. This helps in incorporation that is into broader management models in order to contribute to the boarder scientific work.

Stock assessment primarily focuses on the individual population. But in order to have complete understanding of the population requires to be considered in a broader context. This includes meta-population structures, species interactions which involves predation, competition, etc.

There are variations in conditions in marine environments which includes water temperature, primary productivity, water current changes, etc. When the focus is generally on the exploited populations, the techniques of stock assessment can also be applied to the non-exploited populations.

Stock assessments are basically the result of teamwork. These teams are basically responsible for bringing together scientist that belongs to the variety of disciplines. That can include from the fishing gears expert to biologist, geneticists to statisticians.

Fisherman including harvesting industries and other interested stakeholders also plays important role in data collection, undertaking research, data interpretation, facilitating scientific framework and also in decision making.

With the help of these ways, the stock assessment science gets broaden in modifying practices and also for opening new communication channels that can be used for discussing assessment results.

Stock assessment and researches related to it helps in including a wide variety of initiatives that can be aimed for monitoring fisheries and fish stock. That can evaluate the status of marine resources and also assessing the impact of human activities on these resources. Stock assessment initiatives typically includes:

- some of assessment function includes activities such as analysing data, conducting risk analyses and writing scientific, constructing, and running models of population dynamics and technical reports on resource status;
- Scientific peer review includes a well-defined process to review scientific data, analyses, and to communicate results or conclusions;
- stock assessment focuses on fundamental research that is on biological characteristics, migration, fish growth, recruitment processes, etc. there are also research on new approaches or survey techniques;
- there is monitoring of resources through various survey techniques;
- there is sampling of commercial catches;
- there are advisory functions, which includes providing information on specific questions related to the management of resources. this may be either on demand or through participation in committees or regular meetings of advisory bodies; and
- it also helps in administrative function.

ITQs are used to place demand for the quality in the stock assessments. This basically effects the scientific advisory process. It is therefore made as credible and understandable which is possible through the scientific advisory process.

While maintaining ecosystem perspectives "it is important to focus on the scope of investigation with respect to the biological research". With this is also important to strengthen the science of administration functions.

When a cost recovery policy is a part of the ITQ system, then there will be certainly increase in the interest by the fishing industry which determine the nature of the activities. This is done by setting the research priorities and the stock assessments methods and results.

It is needed to get ensure about the essential long-term monitoring functions. This is also essential for maintain the appropriate balance between generating new knowledge and the provision of timely assessment advice that is required to be maintained.

There is also need to tighten the research management controls and to define the true cost of research and monitoring activities. There will be a requirement of defensible policy on recoverable and non-recoverable research costs.

Therefore, ITQs are affecting the resource of research and analysis function in a number of important ways. There is improvement in the linkage between the research, management, and the fishing industries. Industries will increasingly contribute to the stock assessment and will also participate in the peer reviews of management process.

The science industry co-operations get extended to the provision of vessels that is used for research survey, to conduct complementary industry survey, fishing programs in order to establish the biomass trend indexes, provision to improve the catch, and effort data from the industries. These all help in reducing otherwise expensive survey costs.

2.7.3. Socio-economic and research analysis

Setting of beneficial fisheries policy is as economic as a biological problem. With respect to the strength and influence of the research functions, most of the countries are said to make the substantial investment in understanding the concept which controls their living marine resources through natural science research. They are said to invest less in the research because that would produce sustainable use practices.

There is some basic economic information that are needed for a continuously collection and regularly tabulation are:

- Fishing effort and the use of inputs by fishery and fleet segments and the nature of the fishing fleet and processing capital;
- Costs and earnings of enterprises;
- The industrial structure which includes ownership, vertical, and horizontal integration of the fishing industry, labor, and other contracts, etc.;
- Commodity flows;
- The asset and liability structure of the fisheries companies; and
- Input and output prices.

The above data will be only provided when there is a disclosure which is mandatory. The data is collected by the inputs analysis and modelling is required to evaluate the alternative fisheries management measures. It also provides operational advice in order to set the TACs and to evaluate the biological measures like area, mesh size and to determine rent and cost of recovery levels.

ITQ systems are used to create new demands for the socio-economic analysis. This is so because of the need to understand and evaluate the changes. The changes are related to how the fishery operates the adoption of property rights management approaches.

2.7.4. Operational planning

Operational planning functions in the development of annual harvest plans that are made for fisheries. These plants can be applied to only those who are license holders who have access to them. There are some of the jurisdiction that are using fishery plans without ITQs in their management system.

When there are developing plans on a meaningful scale provides a forum for dialogue. It also negotiates between the license holders, researchers, and the resource managers. This is needed especially where the fisheries are geographically distributed and are also diverse.

When there is the use of ITQs in the fishing plans, they are said to translate the annual TAC/TACC into quotas for the respective participants. They also help in providing an efficient way which can fine-tune other management measures which include gear restriction, fish size, and biologically based closed areas. It also engages the license holders while selecting other appropriate measures that can of self-control.

The use of feedbacks is included to have strategic planning function that are used to secure the need of change to regulatory or policy framework. There are operational plans that are needed to identify sensitivity or vulnerable issues that are needed to be addresses through the targeted enforcement activities.

The operational planning functions are playing an important role in the introduction of the ITQ system. In ITQs system, there is need for a fishing plan. As the planning process can facilitate the collective action on management.

They help in launching other measures such as resource enhancement initiative that are required in coordinated actions. The use of ocean space has become so intense and there is also increase in the competitive users. This has become an increasing consideration while planning the operations of fisheries.

These conditions result in demanding for the control of competing users in order to protect fisheries. There are also some restrictions of fisheries in certain areas that permits the activities of other uses. There are some horizontally integrated planning bodies that are likely to emerge in order to resolve conflicts of use.

Secretariat supports the fishery operational planning. Each of the individual fishery requires the support of 'Fisheries Manager.' In this way a person gets occupies as part time for a small fishery or full time for a complex fishery.

The work of these managers is to coordinate with the input from research, enforcement, fishing industry, and other groups to develop the operational fishing plan. The fishery manager is also responsible for managing the approval of plans and regulations that are needed to implement these plants.

2.7.5. Management of fishing entitlements

Most of the fishery management systems are designed to perform simple administrative tasks which involves the recording of ownership of license and vessels with that maintaining of data on the physical characteristics of the vessel. But in this case property rights registry system works more functionally.

The quotas registry must be able to convert the annual TAC/TACC to the annual catch limits. This limit is set by the weight of the catch for the individual right holders.

Therefore, the registry should be connected to the catch monitoring system because this helps to track the use of quotas held and to support the season leasing and long-term sales of quotas for the right holders. If the ITQ are of long term and secure then they may serve as a collateral in the capital market.

If the rent or cost of recovery are features of the programmes. Registry is considered to play an important role in revenue collection service.

To support these functions, registry must be dynamic and interconnected with the information system that is used to support other management functions. The skills that are required to operate a registry are being administered and commercial.

2.7.6. Catch effort monitoring

In catch management of fisheries, monitoring, and data collection are often considered as a weak link. In many of the systems catches are estimated based on the sampling or other catch projections. Fleet quotas can be estimated caught and it can be fishery closed based on these elements. There would be no incentive to buy quota to cover one's operations if the catch were, in effect, unrestricted.

ITQ system are said to require difficult record-keeping, mandatory reporting and disclosure of rules. Mostly the trip by trip monitoring is advised. This monitoring type includes requirement to report departures from the ports for fishing. This also includes reporting of the time of vessel arrival before the landing of catch and other information related to fishing activities.

This type of monitoring is important in case where there are larger and more sophisticated vessels. In most of the circumstances, the dockside verification of landing by the independent third party is also advised. The number and distribution of the landing ports and the size of the fleet will decide the monitoring cost.

Figure 2.6. *Representation of fish fairs and participation of fishermen in a small amount.*

Source: Image by New York Public Library.

The time of landing catch, port chosen are some of the industry behavior that will also affect these monitoring cost. There are direct forms of cost recovery that are used for monitoring service that can have a significant effect in restraining the cost for these activities.

In ITQ systems high grading and discarding is often defined as a problem. With ITQ system particularly in multi species fisheries, these factors are problem. In order to minimize the practice, the problem should be approached with the combination of drawbacks and incentives.

There should be a general prohibition against discarding of any species at sea. This simplifies the monitoring and the action should be taken at the time when there is a detection of discarding.

This works in combination with the high penalties so as to ensure that all the of the catch is landed and this make easy monitoring and it can be deducted from the annual catch quotas. Flexible trading rules are available those allows fishers to lease or purchase quotas in order to match with the actual catch.

This helps in encouraging compliance with the management plan and it also helps in reducing the need to discard them. At the time of year end, the 'over catch' and the carryover rules that are for the uncaught quota will minimize the amount of incentives to discard in order to match landed catches to the quotas that is owned.

2.8. LIMITED LICENSING AND ITS EFFECTS

2.8.1. Limited licensing as a right of entry

Today there are limited licensing is extensively increasing. Especially where the fish stocks are migratory or linked jurisdiction boundaries. It was supposed that in gear limitations and closures were not correct to deal with the limited licensing problem.

That is why there was a search for a system that can help in reducing the amount of effort, then reducing the time and limit of forms. The simple answer to this is to reduce the number of vessels. The reduction of vessel can be carried out by simply focusing on the license. This innovation or method is to re-interpret the license in the form of entry ticket.

2.8.2. Transition by grandfathering

When it is seen in case of retrospect, the limited licensing regime was not found very different from that in the past. But in the case of transition it is found as a great deal of upset. So, new schemes are seemed to be different when comparing from free and open access.

This scheme was introduced when there was already more amount of fishing. This scheme should be sold politically as a system that can benefit the hard-pressed fisherman. It is important to have new scheme base on the existing fleet and according to the crew.

When the restart takes place, then to start from all over again is not feasible. So, in this case, the allocation of the limited number of the license will help. This process includes various like lottery system, tendering bids, and open auction sale.

When the reservation system gets involved, then the number of licenses is to be held for the specific classes. These classes involve foreigners, owners of vessels, aborigines, and gears, etc. The new licenses were handed to those who claim to be in the recent fishing activity. This is now commonly known as 'grandfathering' approach.

2.8.3. Getting to the right number of licenses

Limited licensing is considered to put no limitations on a vessel catch. But it does not include those who are already imposed by the existing gears and closures regulations. This means that it could not bring the fleet catch down to TAC.

Because of this, the government must think about the supplements in order to limit licensing. They are supposed to come up with a quite a variety. One way in which one can do nothing is when the fisherman gets retire or reaches to death then their license number gets cancelled which directly decreases the number of licenses.

Second way is to make the license transferable which directly allows the market to develop. This also help in buying some back at the market places or at the fairs. The finally way is to buy some vessels with the license attached to it.

2.8.4. Is limited licensing a success?

Limited licensing is not given that much importance and it's not that successful. One thing is to reduce the fleet size in a right level. At the time when these schemes were under discussion, the fisherman's make sure that they are eligible.

Their eligibility shows that in the transition period the actual number of increased active vessels. In the early months the price of the remaining license remains increased. In that case, the remaining fisherman begins to evaluate about how high this will increase. As a result of this many of the fisherman rejects to sell their catch at the time of fairs.

In place of other reasons, this scheme did not reinforce the regulations. The increase in the fishing efforts subjects to the fish stocks exerted by a decreasing number of vessels that is slowing down gradually. Therefore, the catch increases with the young fishes. This does not impact on the total catch nor on the catch per unit effort (CPUE).

Lastly, the fishing cost gets continuously increase. But the remaining vessels had to still compete each other for the catch.

The open periods continue to shorten and the gear regulations continue to increase the time and equipment needed to fill up a vessel. Private cost must be added in order to increase the public cost of enforcing the TAC.

2.8.5. Limited licensing significance

Despite the near failure, there was to hope for. Limited licensing is said to have a very significant features where some are accidental and some are by design. This prepares the regulatory and the fishermen for the next to come.

This created a connection between the fleets and the stocks. In case of large and offshore fisheries, there are vessels and the owners who were found to be directly connected with the single species or stocks or with the fishing grounds. There were some fishermen who wanted to lower the cost and they also wants to improve the efficiency.

Those fishermen are found to think more about the specific fish stock that is being exploited. These fishermen had to be concerned about the fish stock that had been consider as the distinct part of the industry that was harvesting it. This created an incentive for the fishermen that helps to assist the administrator in enforcing the regulations. It was very clear at the time of flouted regulations, who were the fishermen who were hurt.

In order to develop a helpful relationship between fishermen and the regulators, the creation of incentives takes place. The condition has become obvious were the insiders use to held the license and the outsiders were now ignored. The insider has the capacity that help them to develop an interest in conserving and improving the stocks.

When the previous factors such as fierce, perpetual, wrangling, where the regulators are being regarded as imprudent policemen are defied and evaded. These can now be seen in they are needed for the replacement by the joint consultation and for the collection of information.

These opportunities and incentive were the evident. But these evident were hidden by the rivalries, competitive, and fishing behavior of fishermen. It was made clear to many of the fishermen that they can do better individually by getting larger parts of the TAC for their own classes.

Ports and groups of vessels are helping to increase the stock that is needed to be fished by everyone. This process involves the devoting of one time in order to fight with the system rather than using it for their benefits.

Limited licensing has created the incentives of working together were the overwhelming by the incentives to conceal, discourage, evade, and free rides.There are some words that are used to describe the atmosphere in some of the important fisheries are hostility, distrust, and obstruct into which a system is introduced that allows only limited licensing.

2.8.6. Focus on the license

When the fishermen get limited towards the license then the fishing license becomes important in themselves. They have come as a convenient cog in a registration and administration systems. Before the year 1960s, it was very easy to have the ocean fishing license and vessel license. As many of the paperwork was done without dipping the regulatory system into the chaos.

The first most importance in limited license lies in the word licensing. The license is considered as a permit, certificate that helps to give right to the man that specifies that specific person in eligible to fish in the specified area and for the time.

When the person does not have the license, then that person has no such rights. These were established just to provide a convenience of the administrator, but it was considered as a control against other fishermen. Now one can see that the license has given its holders the rights with the new attributes which are discussed below:

2.8.7. Quality of title and security

Now for the license, person must pay. Which means that license is no longer free now. Many of the fishermen are arguing for this and many are forced to pay for them. In order to force the fishermen to take the license seriously, the government has pushed themselves more.

As it had to protect the holder against those who claim that they are better right to a license holder than those who are holder of themselves in order to protect the quality of title. They had to resist with the enticement to cancel them, take them back and even to reduce their flexibilities rather honouring them.

2.8.8. Exclusivity

As there is no longer open access. Exclusivity is when the property holder utilizes and manages the resources in their own property without letting outsiders to interfere. It is now excluded that the rights of those who might fish were far from being perfectly exclusive but they are more exclusive.

2.8.9. Transferability

There were some of the license that were not transferable to other in earlier system. But those who had them, they can be allowed to use them till they die. But after their death the license is being returned to the state. This license system of transferable has become more famous among all the fishermen.

But their license was considered to have some value, but in this case, they cannot cash in on it nether they can pass it to their heirs. Even they cannot use it as a security in case of loans. This way, they agitate to have license made transferable. They instead found the way to evade the prohibition. But soon, nearly every type of license which is limited becomes transferable or merchantable.

These three attributes or characteristics partially complement each other. For example, transferability become more valuable if the license is more secure and with this if they had a longer duration for their license. A good title and security are worthwhile when obtaining only if the rights had some exclusivity.

2.9. CONCLUSIONS

The definition of property rights was best given by Bromley in 1991. He said that 'Property is not an object such as land but is rather a right or group of rights to a benefit stream that is only as secure as the duty of all others to respect the conditions that protect that stream.'

This chapter explained that there are certain property rights that are basically followed in the sea, ocean or the areas used for fishing.

These rights are given to the fishermen, to provide the fishermen all the facilities and to maintain the quotas and availability of the resources.

As there is increase in the demand, with the demand the availability of fishermen also increased. If these fishermen are given freedom to have catch according to them, then there will a lot of wastage. This wastage can directly impact the resource availability.

So, in order to maintain every area with the sufficient number of sources, there are rules that are bought into the society. These rules are given to each fisherman which contain certain factor that are necessary to be kept in mind while fishing.

Some of the rules includes the catch amount, area to do fishing and many other factors. As earlier fishing was done without any license, but now it is compulsory to have the license in order to have fishing.When these property rights were coming in to access, the management of these rights were also important. Certain policies were developed to being an equal right for all the fishermen. But there are still some of the people who are not obeying these rights, as they might belong to some special classes like foreigner, partners of vessels.

Earlier there were no limitation of licensing. Which means that license was given to everyone.

There were no such rules or rights. People can have fishing at any place they want. Even the use of license was not that much. There was no proper checking of these license.

Anyone who carry license, that doesn't depend weather that license belong to that person or not. But now the rights have made this difficult. As now to have license is not an easy task. Person must be eligible for the license.

Before the license was non-transferable, which means that the person holds the license till death and after their death the license was given back to the state. But no, people can transfer their license easily.

Likewise, there many rights that are important to have fishing. But their maintain ace is also very important. With the increasing demand of fishes, the availability and their growth are also important. To have proper availability of these resources, government has planned to provide the fishery business to the individual people for certain lease.

REFERENCES

1. Arnason, R., (2005). Property rights in fisheries: Iceland's experience with ITQs. *Reviews in Fish Biology and Fisheries, 15*(3), 243–264.
2. Arnason, R., (2006). *Property Rights in Fisheries: Iceland's Experience With ITQs*. [online] Springer Link. Available at: https://link.springer.com/article/10.1007%2Fs11160–005–5139–6(accessed on 8 March 2020)
3. Davidse, W., Harmsma, H., Van, W. M., McEwan, L., Vestergaard, N., & Frost, H., (1997). *Property Rights in Fishing Effects on the Industry and Effectiveness for Fishery Management Policy*. [eBook] Available at: https://core.ac.uk/download/pdf/29324226.pdf (accessed on 8 March 2020)
4. Fao.org. (n.d.). *Use of Property Rights in Fisheries Management-Theory and Practice*. [online] Available at: http://www.fao.org/3/x7579e03.htm#b3-Selection%20of%20a%20Property%20Rights%20Management%20System%20L.%20G.%20Anderson (accessed on 8 March 2020)
5. Fao.org. (n.d.). *Use of Property Rights in Fisheries Management-Applications of Rights-Based Management*. [online] Available at: http://www.fao.org/3/x7579e04.htm#APPLICATIONS%20OF%20RIGHTSBASED%20MANAGEMENT (accessed on 8 March 2020)
6. Fao.org. (n.d.). *Use of Property Rights in Fisheries Management-Rights-Based Fisheries Management-General Perspectives*. [online] Available at: http://www.fao.org/3/x7579e05.htm (accessed on 8 March 2020)
7. Hannesson, R., (1994). *Rights Based Fishing the Role of Property Rights in Fisheries Management*. [eBook] Newfoundland. Available at: http://dlc.dlib.indiana.edu/dlc/bitstream/handle/10535/5707/rights%20based%20fishing.pdf?sequence=1 (accessed on 8 March 2020).
8. Property Rights in Fisheries: How Much Can Individual Transferable Quotas Accomplish? (2012). [eBook] Iceland. Available at: http://citeseerx.ist.psu.edu/viewdoc/download?doi=10.1.1.862.2786&rep=rep1&type=pdf (accessed on 8 March 2020).
9. Shotton, R., (1999). Use of Property Rights in Fisheries Management. [eBook] Rome. Available at: https://books.google.co.in/books?id=s5Y1YeqD-iAC&pg=PA155&dq=Market-Based+Fisheries+Management&hl=en&sa=X&ved=0ahUKEwjO9_u7ga3nAhWFzTgGHSrbC6kQ6AEIiQEwCw#v=onepage&q=Market-Based%20Fisheries%20Management&f=false (accessed on 8 March 2020).

3

Current Status of Marine Fishing and its Problems

CONTENTS

This chapter discuses mainly about the status of the fisheries globally. In the first half, it mentions the global overview regarding the marine fishery resources. This half provides insights about the major characteristics as well as considerable changes that is done in the fishery stock.

In the second half, the chapter provides detailed description about the limits of global production including the food -web limitations and degree of fish -stock utilization. It discusses the capture fisheries production in global terms. In the last of the chapter, it provides insights about the sustainability status related to the data-limited fisheries and also the global challenges faced by the fish stock.

3.1. INTRODUCTION

Considering the evolution and further history of human, it can be said that fish has been one of the major sources of food and fishing activity was undertaken from oceans, lakes, and rivers so as to get this food.

Despite the fact that the ocean productivity is unlimited as well as renewable, along with the increased human knowledge on nature as well as its resources, and the dynamic technology which is being developed by the human being in order to exploit these resources, the renewable living aquatic resources seems to be exhaustible.

Globally, the livelihoods and the way of life led by the fishing communities are threatened because they tend to struggle in order to cope up with dwindling fish stocks and the increasing regulatory regime. The major problem is overfishing in global fisheries, which in turn plague 32% of stocks around the world and a massive 72% in the EU.

The present scenario is of global warming as well as climate change and this has brought up fishing sector under its ambit. There have negative implications of stock depletion and less recruitment for food security and that is particularly major threat to the economy and well-being of the coastal communities, job opportunities, food security, income, and livelihoods and also, the traditional cultural identity and specifically to those who depend upon the fish as their main source of animal protein and income.

In the year 2009, the global marine fish production was 80 million tonnes and that was directly providing employment to 34 million people in the fishing operations. Besides this, in the year 2008, there were more than 3 billion people who were supplied with marine fish with at least 15% of their average animal protein intake.

Therefore, it is important to ensure that the fisheries resources will continue in order to provide realistic support and benefits for both present as well as future generations. For this purpose, promotion of responsible fisheries is required in order to maintain the long-term prosperity as well as sustainability of marine fisheries.

There are a few instruments which focuses on restoring the fish stocks at levels which are capable of producing their Maximum Sustainable Yield (MSY) and those are United Nations Convention on the Law of the Sea (UNCLOS), the UNFSA that is United Nations Fish Stocks Agreement (UN, 1995) and the FAO Code of Conduct for Responsible Fisheries (FAO, 1995).In order to achieve the above objectives, it is important for the fishery management and development authorities to undertake assessment of the state of fish stocks and further, develop effective policies as well as management strategies.

In general terms, considering the basis of a single-species, there are a number of marine fisheries which are fully exploited or overexploited, whereas, there are a few which seem to have the potential for increased exploitation.And generally, this is true for both the United States as well as globally, especially in estuarine, nearshore, and continental-shelf fisheries, which in turn produce approximately 75% of the fish catches of the world (Pauly and Christensen, 1995).

There is a primary source of global information about the condition of fisheries and that is the Food and Agricultural Organization (FAO) of the United Nations (UN). In the United States, there is a task of carrying out assessments which has been primarily the responsibility of the NMFS that is the National Marine Fisheries Service of the National Oceanic and Atmospheric Administration, U.S. Department of Commerce.

There is a council that is regional fishery management councils, which was established under the authority of the Magnuson-Stevens Fishery Conservation and Management Act of the year 1976, and they are also involved in the assessment of fish stocks in federal waters.In order to conduct the assessments of migratory fish stocks in state waters, the interstate fishery commissions in the Atlantic, Gulf, and Pacific regions work along with states and the NMFS.Despite the fact that the assessments and statistics from FAO and NMFS provide only an imperfect characterization of the status of global as well as U.S. fisheries, the assessments (which are corroborated by different kinds of evidence) tend to provide a reasonably accurate description of the entire picture.

3.2. GLOBAL OVERVIEW OF MARINE FISHERY RESOURCES

3.2.1. Profile of catches

According to FAO, in the last 60 years, the fisheries and aquaculture sectors of the world have gone through a dramatic development. And also, there have been large increases in their production. In the year 1950, total world fish production was only around 19.3 million tonnes.But this number has increased dramatically to 163 million tonnes in the year 2009. The largest contributor to the world fish production have always been the marine capture fisheries.In the year 1950, the marine captures were around 16.7 million tonnes and furthermore, it accounted for approximately 86% of the total world fish production. Considering the last two decades, it can be said that the marine and inland aquaculture have been expanded rapidly, and the relative contribution of the marine capture fisheries to the growing total world fish production has decreased.However, there have been contribution of marine capture fisheries to the fish production and that was around 49% in the year 2009. And therefore, the marine capture fisheries have been considered to be the largest sector while comparing the Mariculture (21%), freshwater aquaculture (23%), and inland capture fishery (6%).

Figure 3.1. Aquaculture pens in Bay of Luoyuan, China.

Source: Image by Flickr.

There are different development stages which are experienced by marine fisheries further increasing from 16.7 million tonnes in the year 1950 to as high as 87.7 million tonnes in the year 1996. Then it starts decreasing so as to stabilize and became equal to 80 million tonnes along with the interannual fluctuations.

In the year 2009, the global recorded production was around 79.5 million tonnes. In the late 1950s and 1960s and in the period in between 1983 and 1989, there was rapid development that took place.

Mainly by shipbuilding expansion post-war in the period of 1950s, the new technologies like the steam and motor trawlers in the period of 1960s and the extension of the jurisdiction to 12 nautical miles by most of the coastal states, the first boom was believed to be caused. This is the region which encompassed the most productive upwelling and continental shelves of the region (Sanchirico and Willen, 2007; Engelhard, 2008).

The second expansion was rapid and was related with the extension of jurisdictions from 12 to 200 nautical miles along with the establishment of EEZs that is exclusive economic zones under the legal foundation of the UNCLOS (Sanchirico and Willen, 2007). After a peak was reached in the year 1996, there was decrease in the global landings in a gradual manner, that dropped by about 10% by the year 2009.

The variation in catches is reflected by the subsequent fluctuations from a few highly productive areas, specifically in the Northwest Pacific and the Southeast Pacific. These areas account for a large area of landings from pelagic species.

3.3. MAJOR CHARACTERISTICS AND SIGNIFICANT CHANGES

In the abundance of a fish stock, fluctuations are said to be a common phenomenon, specifically for the low-trophic-level species. As a result of this, the catch of this stock also usually oscillates in a much similar manner. Nevertheless, at a regional level, most of the times, the fluctuations in catch appear to be less marked.

This is because of some fact that:

i. Most of the time, the fish stocks within an ecosystem compensate for one another since, they increase as well as decrease in abundance; and

ii. Aggregation over catches of all the species usually soothes out the variations of low-trophic-level as well as short-lived species.

The reason being that the abundance and catch of high-trophic-level and long-lived species most of the time vary less. Considering the three groups, overall catch trends was analyzed. The three groups are Eastern Central and Southwest Atlantic, and Southeast, Northeast, and Eastern Central Pacific, and Northwest Pacific.

They had large fluctuations in landings. In the Southeast Pacific, the largest fluctuation is observed. In the period between the year 1970 and 1973, a drop of around 10 million tonnes occurred.

Then, the fish stocks recovered and further it produced a catch of more than 20 million tonnes in the year 1994 which was the highest catch of all time. From this area then, the catch dropped by 12 million tonnes again in the subsequent 5 years (between the period of 1994 and 1998) and was at around 12 million tonnes in the year 2009.

This was almost as high as its first peak in the year 1970. The large interannual variation in catch from the Southeast Pacific is caused by the large proportion of the pelagic species in catches from the area.

Anchoveta (*Engraulis ringens*), Chilean jack mackerel (*Trachurus murphyi*) and South American pilchard or sardine (*Sardinops sagax*) were the top three species.

Considering it all together, they consider for more than 80% of the current as well as historical catches. In recent decades, they do have altering periods of high as well as low abundance.

In such regions, large catch fluctuations are common and therefore, are mostly a consequence of the periodic climatic events which is referred to as El Niño. Fishing success is affected by El Niño events as well as stock abundance and productivity for longer terms.

There are considerable fluctuations which were also reported for other regions, despite the fact that they have their combined effect on global catches which was less noticeable.

Let's take an example. In the region namely Northeast Pacific, the fish production reached at the highest of 3.6 million tonnes in the year 1987 and further, it decreased to 2.2 million tonnes in the year 2009. After that, it is followed by a slight recovery to 3.2 million tonnes in the year 2005. In the late 1980s, the Northwest Pacific showed an oscillation in between 20 and 24 million tonnes.

These fluctuations were the result caused by catch and, abundance changes that occurred in the Japanese pilchard or sardine (*Sardinops melanostictus*) and Alaska pollock (*Theragra chalcogramma*). Considering the whole, these, and some other changes in the catches and the state resources all have been described. Considering the year 1981, in the region named as Eastern Central Pacific, catches have fluctuated in between 1.2 and 1.8 million tonnes.

In the year 2009, the catch was at a peak of approximately 2 million tonnes and this can be probably because of the recovery of California pilchard or sardine (*Sardinops caeruleus*). Considering the year 2009, they yielded 0.8 million tonnes, which is an all -time record high and a little bit higher in comparison to the high catch of 7,20,000 tonnes in the year 1936, and this occurred during the previous high 'regime' of this species.

This period in which the catches was at its peak lasted from the late 1920s and was there throughout the early 1940s. In the Southeast Atlantic, the total landings have also fluctuated approximately 2 million tonnes considering the period of the late 1980s.

There are species which have showed strong fluctuation in this area named as Argentine hake, Argentine anchovy, Argentine short-fin squid and Argentine red shrimp.

In the landings of the second group (declining landings), there were temporal fluctuations which were weaker.

From the Northeast Atlantic, the landings have continued the decreasing trend which was seen since the period of mid-1970s. This was mainly the result of the decline in Atlantic cod (*Gadus morhua*) since, the period of late 1960s, having a bounce back in the period of 1990s.

It is worth noting that the landings of the blue whiting (*Micromesistius poutassou*) decreased back to below 1 million tonnes in the year 2009, which tends to increase gradually since the period of 1970s and further, reached a peak of approximately 2 million tonnes in the year 2003.

Also, Sandeels (*Ammodytes* spp.) experienced a striking as well as marked drop to below 0.4 million tonnes in the year 2009, after peaking at more than 1.2 million tonnes in the mid-1990s.

In the region named as the Northeast Atlantic, fish production decreased to a low of 2 million tonnes in the year 1994.

This followed the collapse of groundfish stocks off eastern Canada. However, the catch has been about 2 million tonnes since it was established.

In the period of 1970s, the collapse in Atlantic cod and in the early 1990s, the collapse of American plaice took place which was in turn balanced out by the increase in catches of low -trophic -level species like the American sea scallop and American lobster.

There was a period of declining catches which was experienced by the Western Central Atlantic, Mediterranean, and Black Sea and Southwest Pacific, but it was to a lesser extent.

In the last decade, large reductions in the catches have been seen for round sardinella, ocean catfish NEI (not elsewhere included) and requiem sharks NEI in the region of Western Central Atlantic, blue whiting and common octopus in the region of Mediterranean and Black Sea for mullets, and in the region of Southwest Pacific, it was seen for blue grenadier and oreo dories NEI.

There were some signs of declines which was showed for the West Indian Ocean and Western Central pacific which was among the three areas showing a continuously increasing trend in the catch, and this was shown for in the last few years, despite the fact that these may be natural fluctuations.

While examining this trend at the species level, there were large declines which are further clear for skipjack and yellowfin tuna and for the natantian decapods NEI in the West Indian Ocean.

However, there were increase in the catches of other red fishes, Indian oil sardine and giant tiger prawn which in turn balances the decreases in those species. In the similar manner, there were reductions in sharks, rays, skates etc.

NEI and in penaeid shrimps EI from these areas, which was compensated with an increase in skipjack tuna, threadfin breams NEI and natantian decapods NEI in the region around Western Central Pacific.

The region of East Indian Ocean is considered to be the only FAO area which stands out from all the other areas, which has not shown any sign of decline in total catch and no clear, there was decline in catch which has been seen in major fish species.

Species such as tunas and tuna are collectively the most valuable fishery resources which are exploited in the high seas. In the Pacific Ocean, their total production is highest which was followed by the Atlantic and Indian Oceans.

Catches of tuna and some species similar to tuna, increased from less than 0.5 million tonnes in the period of early 1950s to an all-time high of 5.5

million tonnes in the year 2006. Since the year 2003, the catch has stabilized at around 5.4 million tonnes. Among the species, in the year 2009, skipjack tuna (*Katsuwonus pelamis*) accounts for approximately 47%, and at 2.6 million tonnes.

20% of the catch (1.1 million tonnes) was contributed by Yellowfin tuna followed by bigeye tune at 7% (0.4 million tonnes) and kawakawa at 6% (0.3 million tonnes) in the year 2009. Total catches which were from the deep-sea fisheries were recorded to reach at a peak of about 3.6 million tonnes in the year 2004 and then it dropped back to 1.9 million tonnes in the year 2009.

The Atlantic Ocean provides support to the largest deep-sea fishery, which further contributes to about 80% of the total deep-sea catch in between the period of 2000 and 2005. This is followed by the Pacific Ocean and Indian Ocean. This was the dramatic decline, and this can largely be attributed to the decrease in the catches which were reported to be of the blue whiting in the region of Atlantic Ocean.

In the year 2004, the catch of blue whiting decreased from the 2.4 million tonnes to about 65,000 tonnes in the year 2009, and this owes to a decline in recruitment, spawning stock biomass and so, reduced the quotas (ICES, 2011). The species which have yielded a high average catch in the last five years that is from 2005 to 2009 in the region of Atlantic Ocean further including the Patagonian grenadier (*Macruronus magellanicus*), Greenland halibut (*Reinhardtius hippoglossoides*), southern blue whiting (*Micromesistius australis*) and ling (*Molva molva*).

There are top five species which are namely blue whiting (poutassou), which contributes 1.5 million tonnes in 2009 and further, accounts for 35% of the total catch of deep-sea fisheries. This is followed by hairtails and scabbard fishes NEI (1,35,000 tonnes, 7%), Patagonian grenadier (1,32,000 tonnes, 7%), blue grenadier (1,12,000 tonnes, 5%) and Greenland halibut (97,000 tonnes, 5%).

There are some fishes that have experienced a clear decline in catch. For example, Greenland halibut, Patagonia grenadier, southern blue whiting, blue grenadier, orange roughy, and oreo dories. There is some recurring pattern present in some of the areas which is a medium to long term change in the catch composition. This results in the decline of some fish stocks that is being traditionally been dominant.

For example, in some countries like in the Northwest Atlantic, there is an increase in the catch of the species like mollusks and crustaceans which brought a noticeably decline in demersal fishes catch. In Northeast Atlantic there is continues reduction in the catch from the continues decline in Atlantic cod from the late 1960s. But that has been balanced out by the increase in catch of those species that have low value comparatively, such as blue whiting and sandeel fish.

Whereas in the Northwest Pacific region, there is a decline in the catches of some species like Japanese pilchard or sardine and Alaska pollock. These species use become decline when there was increase in the catch of Japanese anchovy (*Engraulis japonicus*), large head hairtail (*Trichiurus lepturus*) and squids (*Todarodes pacificus*) species.

Therefore, the causes for these medium to long term changes in the species composition of the marine commercial catches can be because of the multi folds. There can be many causes for this change. Some causes may include adaptation of the industries and markets towards previous unattractive low values species.

Another reason can be the attractiveness towards new species or specific species and focus on the structure of the marine communities. As nowadays, the target-based fishing is the most popular fishing method used all over the world, this has decreased the use of other species.

At the same many other factors like environmental changes or administration shifts are affecting the long-term abundance. This has affected many wild fish stocks. Although, these effects are confusing and it becomes sometime difficult to differentiate in many cases. These reason scans be successful in the areas where research and monitoring of fish resources and the environmental processes are not well developed.

3.4. LIMITS TO GLOBAL PRODUCTION

There are three kinds of information which suggest that the marine fish catch is near, at or above its peak sustainable level. Those are named as estimates of theoretical limits which are imposed by the available primary production, information related to the degree of utilization of fish populations and information on the catch per ton of fishing vessel.

3.4.1. Food-web limitations

The ocean has the capacity to produce fish which is limited in part by the amount of marine phytoplankton which are produced annually.

Figure 3.2. *Image of marine algae.*

Source: Image by Flickr.

In the phytoplankton concentrations, satellite, and in situ measurements and the in-situ measurement of the nutrients, water temperature, irradiance, and primary production allow the estimates of the primary production of the global ocean and also, the regional estimates.

There have been estimations which is done a number of times regarding an upper limit to the potential fisheries of the ocean. These are estimations are done by applying the knowledge of the amount and location of the global primary production, trophic level of the catch and the transfer efficiency of the biomass among the trophic levels.

Global catch data was used by Pauly and Christensen (1995). They divided that as per the tropic level in order to estimate the flow of the carbon up through the trophic levels of the global marine ecosystems.

They estimated that the transfer of the efficiency taking place between trophic levels was around 10% and further, it concluded that about one-quarter to one-third of total primary production in coastal as well as continental shelf waters is needed in order to provide support recorded landings plus discards.

Fishery landings can be higher from ecosystems along with higher levels of primary productions, especially marine areas which are characterized by fronts, convergence and upwelling areas.

Relationships in between catches and primary production (Nixon, 1988) and between fish production and primary production (Iverson, 1990) was used by Houde and Rutherford (1993) in order to estimate a partitioned global fisheries production for estuaries, coastal zones, and upwelling areas.

According to their estimations, there were a total global fisheries production in those ecosystems of 543 million t, from which 111 million t can be removed as yield.

Such estimations suggest that the landings are either near or beyond their sustainable limit, specifically if the fish production lost as discards and unaccounted mortality are considered.

There is another line of evidence which suggest that the global marine catch might not increase, even by fishing at a progressively lower trophic levels, which is provided by performing analysis on the changes in the mean trophic level of marine fishery landings.

3.4.2. Degree of fish-stock utilization

As per the reports presented by FAO, it can be said that the degree of utilization of global fish stocks, further classifying the fisheries as underexploited, moderately exploited, heavily to fully exploited, overexploited, depleted, and some are recovering.

The largest number of fisheries that is 44% are further classified as heavily to fully exploited. Around 25% of stocks have been fished beyond the sustainable limits (overexploited, depleted, and recovering).

During the period of 1992–1994, for the United States, the picture was very much similar despite the fact that they have slight difference in terminology. Approximately 12% of 275 stock groups were classified as underutilized, 34% as fully utilized, 23% as overutilized and 31% were of unknown status (NMFS, 1996).

There are 191 stock groups, the status of which were known and 82% were fully utilized or overutilized. An example from U.S. can be taken of a striped bass (*Morone saxatilis*) which are formerly overexploited and now, is considered to be recovered stock. Georges Bank haddock (*Melanogrammus aeglefinus*) is the stock that represents a depleted stock.

In global terms, it can be said that some increase in exploitation may be possible considering the 32% of the landed species. But according to Garcia and Newton (1997), considering the past experience, heavily to fully exploited fisheries are likely candidates for future overfishing.

Alverson (1994) demonstrated this assertion and further reported (FAO data) that from 1980 to 1990, there is increase in the number of overexploited fisheries by 250%, whereas there was a decrease in the number of underexploited fisheries by around 75%. There are depleted species which are being replaced in the catch that are done today by the species which were less heavily fished in the past.

Let's take an example, the chub mackerel (*Scomber japonicus*), Atlantic cutlassfish (*Trichiurus lepturus*), Atlantic mackerel (*Scomber scombrus*) and saithe (Atlantic pollock, *Pollachius virens*) is replaced by the Chilean Inca scad (*Trachurus murphyi*), South American pilchard or Chilean sardine (*Sardinops sagax sagax*), Japanese pilchard (*Sardinops sagax melanosticus*) and skipjack tuna (*Katsuwonus pelamis*) in the top 10 species list in between the period 1973 and 1993.

In the United States, more commercially valuable fish were replaced by the skates and dogfish on Georges Bank. There is limitation in this process of depletion of one resource and replacement by another by the number of potentially catchable as well as usable species; depleted species cannot return to previous abundance levels.

There can be another example of Atlantic halibut (*Hipploglossus hippoglossus*) and spring-spawning Icelandic herring (*Clupea harengus*) which have not been recovered from overfishing, despite the fact that they probably would if in case, there is reduction in the mortality (Myers et al., 1997). And also, there are a number of shark populations appear to be declining.

For instance, according to Van der Elst (1979), there have been impact of South African anti shark nets on the local populations of oceanic sharks and the resultant as well as unanticipated reduction in the nearshore bony fishes.

Considering the global scale, as per the conclusion provided by the Manire and Gruber (1990), the sharks were overfished by approximately 30% per year in the water resources of US. They cited the domestic demand for shark meat; wasteful fisheries practices, especially discarded bycatch of sharks; irrational dread and the global demand for shark fins which is increasing and is considered to be the major factors which contributed to the excess of fishing mortality of sharks.

Along with this, there are unexploited fish populations which lived longer and have slow growing and they tend to support high exploitation rates, not like the populations of faster-growing, short-lived species.

Let's take an example, there are five species of the genus Sebastes, which include the Pacific Ocean perch itself (*S. alutus*) and the northern (*S. polyspinus*), sharpchin (*S. zacentrus*), rougheye (*S. aleutianus*), and shortraker (*S. borealis*) rockfishes off the northwestern and Alaskan coasts of the United States and the coast of British Columbia.

These all are slow-growing and were severely overfished, despite the fact that they have now largely recovered (NPFMC, 1997; NMFS, 1996). In the Southern Ocean, the marbled rock cod (*Notothenia rossi*) also has been severely overfished (Kock, 1992).

Figure 3.3. *Ocean perch.*

Source: Image by Flickr.

3.5. CAPTURE FISHERIES PRODUCTION

In the year 2016, the global total capture fisheries production was 90.9 million tonnes, which observes a small decrease in comparison to the previous two years. In 2015, the world total marine catch was 81.2 million tonnes and in the year 2016, it was 79.3 million tonnes.

By Peru and Chile, the catches of anchoveta (*Engraulis ringens*), which are substantial yet highly variable most of the time and that is because of the influence of El Niño, which in turn accounted for 1.1 million tonnes of this decrease.

The decrease in the catches affected around 64% of the 25 top producer countries namely China, United States of America, Russian Federation, Viet Nam, Indonesia, Peru, India, Japan, Malaysia, Norway, Philippines, Chile,

Morocco, Taiwan, Republic of Korea, Mexico, Myanmar, Thailand, Iceland, Ecuador, Spain, Canada, Province of China, Argentina, Denmark, United Kingdom; but only 37% of the remaining 170 countries.

By China, the total marine catches were stable in the year 2016 which is the world's top producer by far, but the inclusion of a progressive catch reduction policy in the national Thirteenth Five-Year Plan for the period of 2016–2020 which is expected to result in considerable decrease in the upcoming years, along with a reduction which is predicted of more than 5 million tonnes by the year 2020.

Again, Alaska pollock (*Theragra chalcogramma*) surpassed anchoveta in the year 2014, as the top species in the year 2016, having the highest catches since the year 1998. However, for the year 29017, preliminary data showed a significant recovery of anchoveta catches.

Figure 3.4. *Alaska pollock: also named as Gadus chalcogrammus.*

Source: Image by Wikimedia Commons.

For the seventh consecutive year, skipjack tuna (*Katsuwonus pelamis*) ranked third. All the most valuable species groups having considerable production-lobsters, gastropods, crabs, and shrimps, with an estimated average value by group of USD 8,800 to USD 3,800 per tonne which is said to be marked as a new catch record in the year 2016.

In the inland water, the total catches were 11.6 million tonnes in the year 2016, and there is an increase of 2.0% in comparison to the previous year and was of 10.5% in comparison to the average of the period from the year 2005 to 2014. In the year 2016, there were 16 countries, from which mostly were from Asia, and these countries produced almost 80% of these catches.

The trend of the inland fisheries production is continuously increasing, and this can be misleading. However, as some of the increase can be attributed to improved reporting as well as assessment at the country level and cannot be wholly because of the increased production.

3.6. SUSTAINABILITY STATUS OF DATA-LIMITED FISHERIES: GLOBAL CHALLENGES FOR SNAPPER AND GROUPER

Snapper as well as grouper are considered to be important fisheries resources, along with high commercial value and so plays an important role in the livelihoods and the food security of a number of local communities worldwide. However, the status of a number of snapper as well as grouper fisheries is still unknown, specifically in the cases of small -scale fisheries considering the developing countries.

On the basis of FAO assessment criteria, there are a number of complementary approaches which were explored, and include determination of the status of snapper as well as grouper fisheries, analysis of landings time-series trends and performing investigation about whether other variables could be used as proxies for fishery status. Around half of these fisheries were classified as overexploited, 30% as non-fully exploited as well as 19% were classified as fully exploited.

The FAO landings data indicated the increase in the number of overexploited fisheries over the years and also, indicated that the majority of these fisheries are in transition between the fully exploited as well as overexploited statuses. The Human Development Index emerged as a potential proxy considering the status of the biomass.

There was multinomial modeling approach which explained around 44% of the variability which was observed in the biomass stock status classification data and further indicated a high level of correspondence in between original as well as estimated status, which makes this approach very attractive for the application to other data-limited fisheries.

In the global food security, marine fisheries play an important role and also, provide job opportunities as well as livelihoods for a number of different coastal communities. Demand for seafood production is expected to increase over the next decade, along with the rise in consumption of animal protein while developing the countries and the rapidly increasing human population (Ye and Cochrane, 2011; Watson et al., 2015).Considering the economic as well as ecological importance of fisheries, evaluation is done in their sustainability on the basis of the best available data which is important for improving the management and conservation (Mora et al., 2009; Ye and Cochrane, 2011; Hilborn and Ovando, 2014).

However, effective management of a number of species is challenged most of the times by the limited quantity as well as quality of information that is available (Anticamara et al., 2011; Carruthers et al., 2014; Chrysafi and Kuparinen, 2016).

Life-history information, species-specific catch data and fishery-independent data are seems to be very expensive and also, time -consuming to obtain (Heyman et al., 2014; Arnold and Heppell, 2015; Robinson et al., 2015), and a large number of stocks, which are commercially exploited, lack quantitative assessments as well as reliable estimates of stock status (Costello et al., 2012; Jardim et al., 2015).

The increasingly sophisticated stock assessment methodologies are not always suited to data-poor fisheries (Bentley, 2014) and are only applied to a limited number of stocks which are intensively studied in the developed countries (Ricard et al., 2012).

In order to address these challenges, the development of alternative methods has been a major topic of research in some of the recent years in order to increase the knowledge about the data-limited fisheries.

Chrysafi and Kuparinen (2016) compiled the assessment methods which are most common for the data-limited fisheries and other recent studies have been presenting some of the new methods for evaluating the status of fisheries (Thorson et al., 2013; Rosenberg, 2014; Dowling et al., 2015; Geromont and Butterworth, 2015; Anderson et al., 2017; Froese et al., 2017; Rosenberg et al., 2017; Carruthers, 2018).

However, there are some of the methods among these which still require some of the basic knowledge of the life -history of the species.

The problems related to the data availability tend to be more prominent in the areas having high species diversity and where the fisheries play an important role in the food security, specifically in the tropical areas as well as countries having low -income in Africa, Asia, Oceania, and the Caribbean (Vasconcellos and Cochrane, 2005).

Most of the time, this is the case for snapper and grouper fisheries, which are valuable with high commercial value as well as high demand in the major international markets (Cawthorn and Mariani, 2017).

These fisheries also provide support to the livelihoods as well as food security to many of the locals, small-scale fishing communities worldwide

(Martinez-Andrade, 2003; Teh et al., 2005; Sadovy de Mitcheson et al., 2013; Heyman et al., 2014; Kittinger et al., 2015).

In the common cases, the snapper refers to the species of the family named as Lutjanidae, whereas grouper is referred to as species of the subfamily Epinephelinae. Despite the fact that they are important, the status of many snapper and grouper fisheries is still not known, specifically in the cases of small -scale fisheries considering some of the developing countries.

Where the reporting system is not present or not sufficient (Robinson et al., 2015; Cawthorn and Mariani, 2017). There was categorization done of the 70% of global snapper and grouper fisheries at a low knowledge outcome, along with only 2% of fisheries at a high knowledge outcome (Amorim et al., 2018).

The life history characteristics of these species such as slow-growing, late-maturing, seasonal spawning aggregations, in combination with the external threats like nursery habitat destruction, use of destructive fishing gear, juvenile extraction and illegal, unreported, and unregulated (IUU) fisheries make them specifically susceptible to overexploitation (Sadovy de Mitcheson and Erisman, 2012; Bailey and Sumaila, 2015; Robinson et al., 2015; Erauskin-Extramiana et al., 2017; Erisman et al., 2017).

As per the most recent assessment which is available from the Red List of Threatened Species of the IUCN that is International Union for Conservation of Nature, 12% of the total global grouper species (Epinephelinae subfamily – 163 species) were classified.

In the threatened categories (Critically Endangered, Endangered, and Vulnerable), 13% were said to be near Threatened and 44% of Least Concern, whereas around 30% of the species were considered as Data Deficient, further preventing adequate evaluation (Sadovy de Mitcheson et al., 2013).

According to the Red List status assessments for snapper species (Lutjanidae family – 88 species), this group of species is dominantly classified as Least Concern (82%), but still, approximately 13% of the species are considered to be listed as Data Deficient (International Union for Conservation of Nature [IUCN], 2018).

Despite the fact that, these results are based on the global species trends, they provide support to the urgent need for a better global evaluation of the sustainability status of these fishing resources.

There are very few fully quantitative stocks assessments which have been conducted on snapper and grouper fisheries and those that have majorly focus on North America (United States Management Areas) (Amorim et al., 2018).

Considering the snapper and grouper fisheries, and for most of the data-limited fisheries of the world, independent survey estimates of fish abundance are not much available. The landings which are reported are most of the time considered main and, in many cases, the only, source of information for those resources (Carruthers, 2018).

There are catch -based methods which have been developed, but these methods do require additional assumptions, information, and also, some methodological procedures in order to estimate the stock status (Rosenberg, 2014), and this has made their application much debated (Anderson et al., 2012; Pauly et al., 2013).

3.6.1. Fisheries status

Around 65 species and 111 species of the Serranidae family are included the fisheries database of the Lutjanidae family, a total of 176 different species from 73 different countries.

These countries represented 95% of the global reported landings (by weight) of snapper and grouper in the year 2015 (FAO, 2017). For approximately 18% of the fisheries, no information specifically for species was found.

In such cases, the fisheries were collected by genera or family. For 62% of the 719 fisheries included in the database (out of which 293 are lutjanids and 426 are serranids), the stock structure is unknown.

The number of fisheries by country that is included in the database was very variable. The countries such as Indonesia, India, Mexico, and the United States, with the highest number of fisheries were included in the database.

3.6.2. Biomass stock status and fishing pressure categories

Approximately 68% of the 719 fisheries in the database (that is equal to 490 fisheries) were classified as 'not evaluated' because of the lack of data. For the remaining 229 snapper and grouper fisheries that had information (32%), on the basis of the application of the stock.

Which includes status classification criteria of FAO, 23% were considered to be non-fully exploited (that is around 53 fisheries), 14% were considered to be fully exploited (that is 33 fisheries), 38% were considered to be overexploited (that is 88 fisheries) and 24% were unknown (that is 55 fisheries).There was a final sample of 174 fisheries having a classification of stock status which was thus obtained by not including the 'not evaluated' fisheries and also, those for which an unknown status was assigned. This final sample represented approximately 24% of the 719 fisheries that are contained in the database.

Considering an exploratory approach, there is no differences which were found in the stock biomass status categories by family, despite the fact that the Serranidae family exhibits a higher percentage of fisheries, not evaluated than the Lutjanidae family.

Figure 3.5. From Serranidae family-Epinephelus tauvin.

Source: Image by Wikimedia Commons.

For the majority of countries, there is a lack of information on fisheries stock status or the status is set as unknown or undefined in some other cases.

Such findings are said to be true across the top 10 producing countries (on the basis of the FAO reported landings) of both groups of species. The top producing countries are namely Indonesia, China, the Philippines, Mexico, Malaysia, Brazil, Pakistan, Nigeria, the United Arab Emirates, and the United States.

There was a large number of fisheries where it was possible to determine the classification of the biomass stock status namely overexploited. But there are always an exception and here there are exceptions such as Madagascar, Sri Lanka and Bangladesh. Here the majority of the fisheries were classified as non-fully exploited and also, Hawaii, where they were mainly said to be fully exploited.Considering the case of United States, Kenya, Puerto Rico, Australia, and Indonesia, only around one-quarter of the fisheries were analyzed to be classified as non-fully exploited or fully exploited.There are a large number of fisheries that is around 89% which lacked information in order to classify the level of fishing pressure. The classification of remaining fisheries (that is equal to 81) was done into the following categories.

These categories are such as no overfishing (around 59%), overfishing (around 31%) and unknown (around 10%). A higher percentage of not evaluated fisheries is contained in the Serranidae family, and this is previously noted for stock status.

At the global scale, the fishing pressure in snapper as well as grouper fisheries is not known. For Tonga, Hawaii, Mexico, the United States, and Brazil, there is larger part of the evaluated fisheries which were classified as no overfishing. Considering the case of the United States and Brazil, overfishing is taking place in some of the fisheries.Despite the fact that India, the Middle East, and Cuba had small size of sample (since just a few fisheries were evaluated), in these areas the fishing pressure status of snapper as well as grouper fisheries suggests that overfishing is occurring.Considering the case of Australia, the fisheries were classified in the category of no overfishing or as undefined, but also, there are a large number of fisheries have no available information at this level.While classifying the biomass stock status of the fisheries which are considered to be non-fully exploited, the level of confidence is said to be low, whereas the fully exploited fisheries have a higher level of confidence. The fisheries which are classified to be overexploited, presented a wider range of confidence from high to low.

The confidence levels for the fishing pressure are comparatively higher, having majority of the fisheries classified at the high-confidence level.

The determination of the fishing pressure classification was done for fewer fisheries in comparison to the biomass status and so, is derived from only one criterion, fishing mortality, which is usually only available for fisheries having a higher level of knowledge.

3.6.3. Biomass and fishing mortality indexes

Considering the fisheries having quantitative information, the Biomass Mortality Indices B_{index} (sample size = 41) and Fishing Mortality Indices F_{index} (sample size 53) were calculated on basis of the most recent estimates of biomass as well as fishing mortality (or proxies) in relation to their reference points.Both groups of species namely Lutjanidae and Serranidae, present average B_{index} values are slightly below zero and this in turn suggests that, generally the biomass of those resources is slightly below their biological reference points. Now, considering F_{index}, both the families show signs of overfishing (here $F_{index} > 0$), despite the fact that the Serranidae family showed a higher dispersion of F_{index} values.

According to FAO, generally, the Western Central Atlantic fisheries are with moderate dispersion, along with two notable exceptions, yellowtail snapper (*Ocyurus chrysurus*) from the northern Gulf of Mexico and Western Central Atlantic and mahogany snapper (*Lutjanus mahogoni*) from Puerto Rico.

In this area, the distribution of the F_{index} values (as per FAO) is positively skewed, having a median below zero, which in turn indicates that no overfishing is taking place entirely but the fact is that some fisheries are pushing the distribution forward.There are fisheries where overfishing take place and these fisheries are vermillion snapper (*Ocyurus chrysurus*) from the northern Gulf of Mexico and Western Central Atlantic and mahogany snapper (*Lutjanus mahogoni*) from Puerto Rico.

Analysis of all the fisheries is done for the Western Central Pacific (according to FAO) and it is seen that the fisheries are overfishes ($B_{index} < 0$), despite the fact that the sample size is very small (around 6).

Considering the Eastern Central Pacific (according to FAO), the sample size was even smaller (that is equal 4); across these four fisheries, the variation of status was very high, but mostly above the non-overfished threshold ($B_{index} > 0$). In the Western Indian Ocean, all of the few fisheries which were analyzed (sample size = 4) are severely overfished.

3.6.4. FAO Landings Trend Analysis

According to the global capture fisheries dataset (1950–2015) provided by FAO (2015), snapper as well as grouper contributed over 6,78,000 tonnes to reported landings, which in turn represents an increase of approximately 37% in relation to the average landings in the previous decade (from 2000 to 2009) and more than double the landings reported in the period of 1980s.

Despite the fact that the total reported landings represented an increasing trend over time, it is important to note that the number of countries reporting has also been increasing through the years, because it has the quality of data reported (e.g., the number of species reported).There are a number of fisheries by decade which have attained their peak value of landings (C_{max})

and also, it is increasing continuously over the decades, except the period that was in between 1980s and 1990s, when the number of fisheries almost stagnated which have reached their C_{max}.The period in between 2000–2009 and 2010–2015 determined the comparison of the exploitation status for the same snapper and grouper fisheries. This, in turn, suggests that all fisheries classified as non-fully exploited status and in the first period, almost half of the fully exploited fisheries shifted to overexploited status. Only a small number of the overexploited fisheries shifted back to the fully exploited fisheries.

For the recent period of data available that is between 2010–2015, the classification of the exploitation status of snapper and grouper fisheries suggests that the Serranidae family has a higher percentage of overexploited fisheries than does the Lutjanidae.Higher percentages of overexploited fisheries are observed in the Americas (in Western Central Atlantic) and in Europe (Mediterranean). Despite the fact that it has fewer snapper as well as grouper fisheries reported, the Eastern Indian Ocean contained a higher percentage of fully exploited fisheries.

3.7. CONCLUSION

Fisheries play a very crucial role because it is said to be a major source of food for millions of people in Asia and the Pacific since time immemorial. The inland capture fisheries as well as freshwater aquaculture provide food security to people who live in rural areas. Whereas there are people who live along the coasts harvest marine fish, shellfish, and seaweeds in order to supply to the nearby markets.There were some early records of fisheries, such as, those of Siam in the year 1925, which in turn stated with confidence that the aquatic resources were abundant as well as varied. Also, they were adapted to all the requirements of the people so it is important not be necessarily depend on any other countries for supplies of aquatic food.

There is richness of the resources which in turn emphasizes on the activity of the force of nature while replenishing the supply and also, it provides its support to the belief that in restoring depleted waters and in safeguarding the future, it can proved to be sufficient for invoking the minimum degree of restriction.There are certain major goals of this work which include providing an overview of the current status and trends of these resources and finding alternative sources of information which could be used in order to determine the status of snapper as well as grouper fisheries, along with other data-limited fisheries.

REFERENCES

1. Amorim, P., Sousa, P., Jardim, E., & Menezes, G., (2019). Sustainability status of data-limited fisheries: Global Challenges for snapper and grouper. *Frontiers in Marine Science*, [online] Vol. 6. Available at: https://www.frontiersin.org/articles/10.3389/fmars.2019.00654/full (accessed on 8 March 2020).
2. Review of the state of world marine fishery resources, (2011). [eBook] p.3. Available at: http://www.fao.org/3/i2389e/i2389e.pdf (accessed on 8 March 2020).
3. Singh, T., & Patnaik, S., (2014). *Marine Fisheries; Its Current Status, Sustainable Management and Socio-Economic Status of the Marine Fishers of Odisha, Through Indian Marine Policy: A Case Study.* [eBook] Bhuwaneshwar, p. 1. Available at: http://www.isca.in/AVFS/Archive/v2/i7/3.ISCA-RJAVFS-2014–022.pdf (accessed on 8 March 2020).
4. The National Academies Press, (1999). Sustaining Marine Fisheries. [online] Available at: https://www.nap.edu/read/6032/chapter/4#29 (accessed on 8 March 2020).
5. www.fao.org. (2018). Sofia 2018-State of Fisheries and Aquaculture in the World 2018. [online] Available at: http://www.fao.org/state-of-fisheries-aquaculture (accessed on 8 March 2020).

4

Eco-Labeling and its Importance

CONTENTS

The chapter of eco-labeling and its importance, mentions about the basic significance of the eco-labeling in the context of fisheries industry. This chapter also explains the basic fundamentals of eco-labeling. This chapter also explains the eco-labeling programs as well as several types of the eco-labeling. This chapter also provides brief about the economics of the eco-labeling.

This chapter also addresses that how the process of eco-labeling of seafood products is useful and significant in order to decrease the environmental damage. This chapter also explains the process as well as opportunities of eco-labeling in emerging nations.

This chapter also provide highlight on the importance of the eco-label such as system certification, certification of the products and services, laboratory services, calibration services, and staff certification.

This chapter also addresses the several advantages and benefits of eco-labeling such as informing the choice of consumer, promoting economic efficiency, simulating market development, encouraging the continuous improvement, promoting the certification, and assisting in monitoring. Also provide highlights on the future trends in eco-labeling.

4.1. INTRODUCTION

There is an extensive agreement in the international community that several numbers of the commercial fisheries all across the world are in distress. The schemes in the context of the eco-labeling are more and more perceived as a simultaneously approach in order to maintain the economic value and the effectiveness of the fisheries, while offering the incentives for the developed or enhanced fisheries management and the conservation of the marine biodiversity.

With respect to the fisheries sector, in the recent interval of time, several numbers of eco-labeling initiatives have been developed in order to compliment and provide assistance to execute the sustainable fisheries management systems. These kinds of schemes have met with the various numbers of level of acceptance.

Standards of the eco-labeling that produces the concerns, which are consisting of the scientific basis of the criteria that are anticipated with respect to the schemes of eco-labeling, the attitude of the customers with respect to these kinds of schemes, and the potential impact of these kinds of schemes on international trade in fish and fishery products.

More specifically, there are various concerns regarding the schemes of eco-labeling that can produce discrimination against the export industries of fisheries from developing nations and nations with economies in transition.

The prime objective of this short introduction is for the consideration of policymakers, civil society and private industries (e.g., non-governmental organizations (NGOs), or community-based organizations and organizations of fisheries) in nations who are having a strong interest in eco-labeling of fish and fisheries product.It is not intended to be an inclusive overview of all the factors and the technical contemplations of the eco-labeling. As an alternative, the main objective is to clarify the major issues that are raised with the help of eco-labeling and recommends.

The significance and the advantages of the stronger engagement with the help of all nations and the stakeholders in the international discussion with respect to the several numbers of possible methodologies in order to address them.

The rationales for engagement are consisting of, below mentioned points:

1. There is a high probability that eco-labeling schemes with respect to the fisheries products are here to stay. With respect to on the experience in other sectors, an individual can expect that these kinds of initiatives will succeed in carving out a niche for their products and that rival schemes may develop.

Even if some of them fails, there will likely be renewed efforts from the end of governments, NGOs (non-governmental organizations) and private industry in order to initiate new ones.

2. Nations that have clear interests in safeguarding their trading interests and ensuring that eco-labeling schemes do not constitute disguised protectionism.
3. More specifically, with respect to the developing nations, the engagement provides the opportunity to promote the initiatives of eco-labeling that are accompanied with the help of adequate financial as well as technical resources.

In order to offset or compensate some of the costs they may incur in efforts to enhance the fisheries management and related execution of the international policies and agreement.

4. Processes like the recent discussions in the Food and Agriculture Organization (FAO) fora offers an opportunity for all the

stakeholders in order to make sure that their varied interests are heard and promoted.

By engaging in ongoing discussions countries can make sure that the present or existing and future eco-labeling schemes conform to basic principles with respect to the transparency and participation and that the criteria for eco-labels are developed with respect to their particular settings and requirements.

4.2. WHAT ARE ECO-LABELS?

Figure 4.1. An illustration of eco-labeling of seafood.

Source: Image by Pxfuel.

The main objective of the initiatives of eco-labeling is to endorse sustainably managed fisheries and highlight their products to the consumers. Product claims that are linked with eco-labeling aim at tapping the increment in the public demand for environmentally preferable products.

Eco-labels are typically relying on the life-cycle evaluation in order to find out the environmental influences of a product "from cradle to grave." Conventionally, the claims appearing on a product must be heads with the help of a chain of custody exercise that documents that the product was derived from, for example, a fishery certified as being "sustainably managed."

Prior to the certification, a bunch of standards or criteria of sustainability against which a fishery is to be assessed, must be created. Finding out and accomplishing the "sustainability" with respect to the fisheries is a very complicated process.

> Eco-labels are seals of approval that are given to products that are deemed to have fewer influences on the environment as compared to the functionally as well as competitively identical products. The rationale for basic labeling information at the point of sale is that it links fisheries products to their production process.

The credibility and the acceptance of the standards is closely associated to how the standards were developed, the standards themselves, and the endorsing or certifying process with the help of which the organizations are assessed against the standard.

The eco-labeling programs usually fall into one of the following categories:

1. The labelling scheme of the first party: These are developed with the help of individual companies that are based on their own standards of the product.

The standards might be based on the criteria that are associated to the specific environmental issues known to informed consumers with the help of the media or advertising. This type of eco-labeling can also be referred to as 'self-declaration.'

2. The labelling scheme of the second party: These are developed with the help of industry associations for their products of the members.

The members explain the criteria of certification, sometimes with the help of drawing upon external expertise from academia as well as environmental organizations. Verification of compliance is accomplished with the help of internal certification procedures within the industry, or employment of external certifying companies.

3. The labelling scheme of the third party: These are generally developed with the help of a private initiator independent from the producers, distributors, and sellers of the labelled products.

Products supplied with the help of the organizations or resources that are certified are then labeled with the information to the consumers that the product was produced in an 'environmentally friendly' fashion.

The label (seal) is generally licensed to a producer and may appear on or accompany a product derived from a certified fishery or producer. Producers are generally anticipated in order to track the 'chain of custody' of their products in order to make sure that the products that are derived from the certified fishery are, in the matter of fact, those that are so labelled. In some cases, the private initiator accredits other organizations to be the certifier.

An accrediting body offers some level of assurance that the certifier has been trained with the help of an accredited training program and is qualified to perform an evaluation against a specific set of criteria in a given field.

On the other hand, the criteria may be developed with the help of a negotiation process between the several numbers of interested parties, they are every so often motivated with the help of the environmental objectives of the private initiators of these kind of schemes.

Environmental organizations as well as the consumers are usually preferred eco-labelling schemes of this kind because of the heightened confidence that private commercial interests will not compromise the criteria applied to the schemes and strict compliance with them based on verifiable and impartial certification procedures.

Eco-labelling systems can be either compulsory or voluntary. Mandatory eco-labels are government-backed and could play as a trade restriction with the help of foreign producers (that is, imports can be rejected if they do not fulfil).

Imports of the products that do not full fill the requirements with respect to the voluntary are not constrained.

In the context of voluntary labels, it is all depends on the manufacturer in order to decide whether or not to apply for the certification of the product, and the choice of the customers whether to buy (or import) an eco-labeled product.

Programs of voluntary eco-labeling may be funded and controlled with the help of private sector. Some, never the less, are funded and sponsored with the help of government.

4.3. ECO-LABELING PROGRAMS

An eco-labeling program is a system which is being used in order to develop a market-based incentive to boost that can validate and establish that are being harvested in a manner that are ecologically suitable.

The incentive is created in the marketplace with the help of the selective purchasing power of consumers, who preferentially purchase the products that are marked with the eco-label, and perhaps pay a higher price for the eco-labeled product.

This offers the products that are eco-labeled as well as the seller with a market advantage over the non-eco-labeled products. The increment in the price, or possibly the increment in the volume of the sales of the eco-labeled product, more specifically, the producers of the more ecologically sustainable products over those producing the products without eco-labels.

It could be possible, that is decreases the sales and returns to producers of less sustainable products, decreases the value or market value of the products that are non-eco-labeled, as well as it produces an incentive for the producers in order to change the harvesting or farming practices to be more ecologically sustainable and enhance their environmental practices.

An eco-label is a logo, a mark, a label or an endorsement of a product that is used as affix to a seafood product at the point of sale that shows or depicts to a purchaser that the product has been produced with the help of procedures that are ecologically sustainable, and is form a source that is well managed.Eco-labels are generally applied to each individual seafood product in order to provide endorsement that is visible at the point of retail sale.The endorsement product from an eco-label is designed, in order to convey to the consumer, the simple message that they can assuredly purchase the labeled product in preference to an un-labeled product.

If they wish to provide their assistance and support to seafood produced in ways that have less ecological impact on the fish stocks and the environment.

The direct implications are that, these kids of products are more ecologically sustainable in nature. An eco-label may be used to a product after it has been certified as being in compliance with the criteria and the rules of an eco-labeling program.

Certification is an outcome of a process of evaluation that makes sure (or verifies) that a product obeys or fulfills with the sustainability standard and a bunch of principles that are developed with the help of incentive program.

A certification of submission can be utilized as a part of eco-label program, and certainly a certificate may be issued, but not all certification systems takes into the direction towards the award of eco-label.

On the other hand, some of the certification systems may not have any direct relationship with respect to the retail marketing problems, and can be used for purely industry or regulatory purposes (e.g., in order to establish compliance with needs of the government for safe food-processing procedures).

In this way, through the course, eco-labeling is normally based on some kind of assessment process and a consequent certification, the process of the assessment process and a consequent certification, the process of the assessment takes into the direction towards a certification of compliance does not essentially always lead to an eco-label.

The most significant difference between the eco-labeling and certification is the form of market-based incentive, that entice the customers in order to purchase with the help of influence of the eco-label at the point of retail sale and influencing purchasing patterns with the help of provision of other forms of the product endorsement or buying the recommendation.

Buying guides and the rating are closely associated to the eco-labels, and on the other hand, they establish a different types of product endorsement, such as eco-labels they also act to entice the purchasing patterns of the customers and are designed to produce incentives that are market based.

Buying guides and ratings that provide information about a product to the customers or to the resellers with the help of adverts and awareness programs of seafood, with the help of these types of tools as wallet or purse cards, websites, and brochures, where "buy" or "don't buy" advice is generally provided.

On the other hand, these products recommendations are not normally affixed to the actual seafood product at the point of sale (and may not be assisted with the help of producers or resellers of these kinds of products), these guides and ratings systems, never the less, they are also designed in order to impact the choice of the customers of the products with respect to the ecological as well as problems with environmental sustainability.

Buying guides and ratings are not essentially based on the results of the certification system, or even on a clear assessment process. Most importantly the first design as visible product endorsements, in some instances is the exposure of the underpinning decision process may not well think out to be important, with the guide standing mainly on the credibility and the reputation of the recommending organizations.

With respect of some systems, the buying recommendation can be used on a formal as well as extensive underpinning decision process. Seafood recommendations with respect to the customers that are based on the environmental as well as ecological considerations were created in the time interval of middle 1990s, and the first seafood recommendation card for the consumer use was issued in the year 1999 in the magazine of US Audubon Society.

Programs that are related with eco-labeling are the voluntary instruments, in which aquaculture ventures like fisheries can opt to submit their products for the compliance evaluation in order to find out, if their products can carry the eco-label.

The voluntary nature of the eco-labeling is a very significant characteristic in the context of the appeal of the consumer, because this concludes that the products that do not require the sustainability standard would not be submitted with the help of the producers for the assessment or the evaluation, and in this way, they do not carry the eco-label.

Consumers can understand this as a meaning that only the seafood products that are eco-labeled are available in their markets or super-markets, are certainly produced in the sustainable manner.

Never the less, this implication is only true if all kinds of seafood products have the similar access to the voluntary process of evaluation and certification, and if the process of the compliance assessment is complete or inclusive as well as is utilized uniformly across the all products that have been submitted to the evaluation or assessment.

This is not always the case, and there are several reasons behind this. First is the process of evaluation or assessment and evaluation that could be costly in nature and preclude smaller or less profitable fisheries from taking participation in several numbers of eco-labeling programs.

The second reason is that there are several numbers of various eco-labeling programs that have various approaches of finding out if a product fulfills the requirement with thc standards of the eco-label and the standards themselves may fluctuate between programs.

4.4. TYPES OF ECO-LABELS

In generally manner, eco-labels are one part of a family of environmental labeling systems, each of which may, as an outcome, in the certification of a product. The International Organization for Standardization (ISO) has defined in three main kinds of environmentally labeling, and eco-labels generally fall into the category of the ISO Type I labels (table 4.1).

According to the standards of the ISO classification system (table 4.1.), it does not provide, or execute, specific performance standards, but as an alternative it provides a uniform frame for work within which a systematic system of labeling can be created and executed within a sector of industry.

Never the less, eco-labeling programs of seafood are based on the standards of the performance, and products can be categorized on the basis of compliance with the standards, as well as on the processes that are used in order to setup the performance standards and in order to find out the compliance of the products with the standards.

This offers an extensive use of classification of the seafood eco-labeling programs into three main types (Deere, 1999) that are based on the characteristics of the sustainability standard and the compliance assessment process:

1. First-party labeling schemes:

These are generally developed with the help of individual producers or resellers are based on their own standards of the products, and can cover the criteria that is related to the particular types of problems that are related to the environment, quality of food and health.

This type of eco-labeling is generally 'self-declaration,' and can be well thought out to be an ISO Type II environmental labeling system.

2. Second-party labeling schemes:

These are generally developed with the help of industry associations for the product of members, and the standard are found out by the organization. Verification of compliance is normally directed with the help of procedures that are internal to the industry or association, or with the help of external certifiers, and may be either ISO Type I and Type III.

3. Third-party labeling schemes:

These are typically developed with the help of the organization external to the appropriate sector of the industry, and in this way, carry a perceived level of independence.

The owner of the labeling scheme generally sets the criteria and awards a label to the product that are independently verified (with the help of a certification process) to fulfill the requirements or the criteria.

Third-party schemes are generally well thought out to be most robust kind of eco-labeling, due to the independence of the criteria and the verification process from the commercial influences, and are generally of the ISO Type I form.

The main theme that these three eco-labeling systems clutch in common is the expressed purpose or meaning and the common significance of developing the market demand with respect to the products that are, linked to the other products, caught, grown or prepared for the consumption in a more different approach which is more environmentally friendly.

The main difference between the three types of the schemes that are related to the standards and the criteria which are used in order to seek out if a product fulfills, obeys, and the independence and the robustness of the process of verification of the compliance, that is finding out that the product correctly complies with the standard and the criteria.

Table 4.1. ISO Categories of Environmental Labeling Systems

Type I	Voluntary, multicriteria third-party programs that award environmental labels to products meeting a set of predetermined requirements	Environmental labels and declarations – Type I environmental labeling – principles and procedures (ISO 14024: ISO 1999a)
Type II	Self-declared environmental claims made by manufacturers, importers, distributors, retailers or anyone else likely to benefit from such claims	Environmental labels and declarations -self-declared environmental claims (Type II environmental labeling)(ISO 14021: ISO 1999b)
Type III	Declarations that provide quantified environmental information about products, using predetermined parameters based on the ISO life-cycle assessment series of standards. Additional environmental information, such as the impact on biodiversity, hazard, and risk assessment on human health and the environment may be given. May be verified by a third party	Environmental labels and declarations – Type III environmental declarations – principles and procedures (ISO 14025: ISO 2006)

Source: Table by FAO.org.

There are more and more levels of independence from the commercial impacts in the standard-setting and in the verification of the product from the first to the second and the third-party systems, but there is nothing in this classification of the systems to suppose that a third-party system of compliance assessment essentially means that the environmental standard is much higher as compared to the first-party system.

Never the less, first-party systems are generally set low standards that can be accomplished very easily, and with respect to the consumers, the gradation from the first-party schemes to the third-party schemes concludes the increment in the level of sustainability, robustness of the compliance assessment and the dependability of the product endorsement as expressed in the product eco-label or endorsement.

First-party certification systems which are consisting of the development of the standards and the criteria within an organization, with a compliance assessment procedure carried out internally (e.g., by a company-appointed assessor).

These kinds of internal assessments are generally carried out in order to evaluate the compliance to a standard and criteria that have been created and internally decided through an organization as being relevant with respect to the requirements of an eco-label for the products that are being evaluated.

These kinds of internal standards can reflect the particular types of problems that are related to the sustainability, but they would normally be set at a degree of performance that would permit the venture in order to comply with the standards most of the time, on the other hand, the application of the eco-label would put some of the products of the venture at a disadvantage, and possibly with the meaningful commercial consequences.

With respect to the industry of the seafood, these kinds of internal standard, on the other hand, fulfilling the requirement and the criteria of the present-day commercial operational, may not essentially be in accord with either government or community opportunities for the environmental influences of the fisheries or the aquaculture operations.In addition to this, since the process of directing the compliance assessment is not generally exposed to either peer review or public critique in first-party certification systems, the compliance evaluation process is not exposed to the independent verification.These kinds of systems are generally well thought out to be a self-assessed system (ISO Type II labeling), and are not well-known for their robustness with respect to accomplishing the most important developments and enhancements in the ecological sustainability.

The prime significance and value of these kinds of systems is to provide an internal audit mechanism for decreasing the gap between the best- and worst-performing products or ventures with respect to the sustainability.

Without drawing a lot of market attention to the problems, which can convey important opposing the consequences for poorly performing individual operators or aberrant ventures and for the sector as a whole.

4.5. ECONOMICS OF ECO LABELING

In the present interval of time, the eco-label landscape is fragmented due to the presence of the several numbers of labels with different standards and is a pre-competitive environment for the organizations in order to play with the several numbers of the unanswered questions.

Figure 4.2.*Economics of the eco-labeling.*

Source: Image by Pxfuel.

In the context of eco-labels, there are various types of standards that are built around, and also, they are having various impacts on the costs and profits, credibility, and the legitimacy, and the ability in order to accomplish the objectives or aims of the eco-labeling.

Concentrating on the standards of the eco-label and with the help of hoteling-type horizontal differentiation model, several issues have been identified with respect to the problems that are standard-specific and characterize primary barriers to the eco-labeling.

Some of the most important findings are mentioned and explained below:

1. More rigorous environmental standards that are enforced with the help of higher-integrity labels do not essentially translate into higher selling prices;
2. Higher prices commanded through the labeled products do not makes sure that a firm will derive higher profits from eco-labeling;

3. Auditing fees paid per product unit being inspected for ensuring compliance with an eco-label standard (instead of, participation fees paid up front) is the primary de facto barrier to business involvement in eco-labeling; and
4. The willingness of the customer in order to pay price premiums for eco-labeled products is not an appropriate setting in order to generate a premium in the market.

A developed literature in the environmental economics points to the potential for the information in order to provide the assistance to regulate the externalities.

For example, some have argued that publicly available information can encourage the organizations in order to become more environmentally friendly due to the market pressure (Arora and Cason, 1996, 1999; Konar and Cohen, 1997, 2001). Certainly, Tietenberg (1998) refers to this possibility as the "third wave" of pollution control.

There is lots of proofs that some of the customers would be willing to pay a sum of amount in order to "protect the environment" (Amacher et al., 2004; Bjorner et al., 2004; Cairncross, 1992; Cason and Gangadharan, 2001; Haji-Gazali and Simula, 1997; Levin, 1990; Wasik, 1996; Winterhalter and Cassels, 1993).

Organizations that utilizes the environmentally friendly production approaches would like to invest on this demand, but they cope up with an issue which is related to the asymmetric information.

Generally speaking, consumers cannot tell the kind of process of the production that has particularly being used by an organization, in this way, they cannot find out when it is environmentally friendly.

Since the environmentally friendly technique is generally more costly, firms would be opposed when opting these kinds of process, which as an outcome, produces larger level of pollution.

One possibly remedies for this informational asymmetry is for the organizations in order to make of "eco-labeling." With respect to the eco-labeling, a third party, either some kind of governmental agency, or a NGO that certifies a product of vendor as an outcome from a process more environmentally friendly.

In the time span of the last 10 years or so, the process of eco-label has developed in an extensive range of nations (Karl and Orwatt, 2000; OECD, 1977; Vossenaar, 1997).

Some of these programs that are related to the certification have become quite popular, as with the German "blue Angel," Japanese "Eco-mark," Swedish "Environmental Choice," and "Nordic Swan" programs (OECD, 1977).These eco-labels are every so often applied to products where the consumers would typically be individually unable to find out the environmental friendliness of the product, for instance, the biodegradability of a paper product, or of the production process itself.In the present interval of time, several numbers of the eco-labeling programs are in operation consider production-related criteria in their assessments of the firms that finds out for the certification. Much of the existing literature that examines eco-labels assumes perfect (as opposed to probabilistic) certification, either explicit or implicitly. On the other hand, it is very fascinating to regard this certification as absolute, this only makes sense if the third party can perfectly determine the compliance with the eco-label's avowed standards at a reasonable cost.

Certainly, in the present interval of time, several numbers of the eco-labeling programs that are provided with the help of a third party, the compliance of the organization with the environmentally friendly process is measured with the help of random monitoring. But when monitoring is done randomly, the certification must be viewed as noisy.On the other hand, the certifying organization might wish to utilize the absolute standards, in practice, this is not feasible. The certifying party cannot be certain that the organization always utilizes the approaches that are environmentally friendly, nor that monitoring the schemes is able to identify any kind of violation perfectly.In addition to this, there are considerable confusions that the standards are correlated with the "environmental friendliness" in a perfect manner (Arda, 1997; Morris, 1997). These points are need to noted or observed, it seems reasonable in order to expect that environmentally friendly organizations would be more likely to acquire as compared to the organizations that are environmentally unfriendly.

4.6. ECO-LABELING OF SEAFOOD PRODUCTS TO REDUCE ENVIRONMENTAL DAMAGE

> Eco-labeling of seafood products has been introduced to address the environmental impacts of fisheries and fish products. A recent study has compared various types of eco-labeling schemes to assess how they could reduce the ecological impact of fishing, offering suggestions for criteria that seafood products must meet in order to receive eco-labels.

Eco-labels are a form of certification that helps consumers identify which products and services are more environmentally-friendly.

They are designed to encourage manufacturers and food producers to develop environmentally sustainable products. Never the less, different schemes have different criteria that seafood products must meet in order to be awarded with an eco-label.

The researchers examined eco-labels for seafood from four different licensing bodies. These were:

1. 'The Dolphin Safe Tuna' -introduced as a 'single attribute' label (focusing on one important issue) to protect dolphins from being caught as a by-product of tuna fishing;
2. The Marine Stewardship Council (MSC) 2 label -a 'multiple attribute' label (focusing on more than one issue) that promotes sustainable fishing by limiting over-fishing and reducing impacts on the marine ecosystem;
3. The Swedish KRAV3 label -a multiple attribute label that encompasses many types of environmental aspects in several stages of the product's life cycle; and
4. The Danish DSLS4 label -a multiple attribute label with a focus on promoting energy-efficient fishing methods with minimal damage to the seabed. This label is presently not in use.

Figure 4.3. Use of eco-labeling of seafood products in order to reduce environmental damage.

Source: Image by Pixabay.

All the schemes are designed to manage aspects of environmental damage from a range of fishing-related activities, but vary considerably in their scope. Although the Swedish and Danish labels are only national accreditations.

Both the systems address several types of environmental impacts including the contribution of fishing to global warming (via energy use) and cover more stages of the seafood products' lifecycle than the Dolphin and MSC schemes, which concentrate on the effects on marine resources at the fishing stage.

The study emphasizes that the environmental impact of seafood products extends far beyond the initial impact of fishing on the marine environment. It makes suggestions for criteria to be included in eco-labeling schemes which could improve the overall environmental record of fisheries.

For example, the carbon footprint of the seafood industry should be lowered and the use of toxic pollutants, such as anti-fouling chemicals used to coat the hulls of boats, should be reduced.One way of reducing energy use would be to promote energy efficient fishing equipment and restrict less efficient equipment, for example, the beam trawl, used to fish along the sea floor.Additionally, a maximum fuel quota or tradable carbon credit scheme could be introduced, as used by many other industries. The latter system would reward energy efficient fisheries by allowing them to sell their excess credits, although managing this could be complex.

Naturally, protection of fish stocks through sustainable harvesting, minimized by-catches and reduced damage to the sea floor should continue to be included in eco-labeling criteria.

The study also makes suggestions for additional environmental criteria that could be added to the popular MSC label, which does not account for the carbon footprint of the fishing industry. In life cycle assessments (LCAs) of seafood products, the fishing stage often produces the most greenhouse gases (GHGs).

However, for some fish products, such as canned mackerel and pickled herring, the processing stage can have the greatest impact in terms of GHG emissions, due to the packaging used.

4.7. ECO-LABELING AND DEVELOPING COUNTRIES

The schemes of the eco-labeling, and more specifically, those that extend the principles of the eco-labeling from the household cleaning goods to agricultural and products that are based on the natural resources have provoked very considerably, which develops the concerns between few nations, more particularly in the emerging nations.Up to the present interval of time, there is no conclusive proof that the schemes of the eco-labeling for other natural resources, for example, forestry products, have on average are proven harmful and damaging to the emerging nations.

With respect to the sector of fisheries, the emerging nations already have the concerns about the influence on their competitiveness of the rules that are related to the fish health and technical standards, fish additives and food safety.

Their concerns are that the schemes of eco-labeling, with respect to the importing nations could simply add to the lair of restrictions and difficult tasks they cope up with. Along with that, numerous opportunities can also be articulated.

4.7.1. Opportunities

Several numbers of groups of the industries, civil society organizations and governments understood the economic as well as the ecological chances that eco-labeling could offer.

4.7.1.1. Environmental opportunities

Several numbers of the governments and industry groups recognize that eco-labeling could offer the required economic profits for better long-term stewardship and availability of natural resources are meaningful with respect to the national economic welfare.

The schemes of the eco-labeling can provide numbers of nations a tool, in order to help them fulfill the commitments that are made under the international agreements in the context of the significant environmental imperatives, for example, as responsible fisheries and the conservation and sustainable utilization of the biological varieties.

The basic rationale with respect to the eco-labeling is, after all to develop political assistance in order to enhance the management of the environmental standards and in order to increase the environmental standards with the help of the choice of the consumers.

4.7.1.2. Economic opportunities

Labeling offer one of the least-coercive market-based mechanisms in order to enhance the conservation of the outcomes. Private sector interest in eco-labeling for fisheries products in both emerged as well as emerging nations is growing, more particularly given the business and export opportunities eco-labeling has generated in some other sectors.

In addition to this, as already noted, it is the potential for growth in the market share of the eco-labeled products that makes eco-labeling a convincing business choice. If the fisheries management develops in response to efforts to obey or fulfill the criteria of the certification, the potential benefits with respect to the fisheries in both industrial as well as developing nations could go far beyond higher incomes which eco-labeled products may generate.

Some people see the eco-labeling as an important element for gaining access to new premium green markets (e.g., market access). For those producers who are willing and presently or potentially have the ability in order to fulfil the sustainability requirements.

Eco-labeling shows an opportunity to add value to the existing products, expand reach in the markets that exists at present interval of time, or maintain the market share in a competitive market.Product differentiation could be a way for some of the exporters in order to enhance or increase their earnings from exports and eco-labels could be one source of these kinds of product differentiations.

There are also hopes that eco-labeling could offer a new chance for attracting capital financial inputs and joint ventures in the countries which are developing at present interval of time.

For instance, some of the emerging nations are hoping to increase their chances at fulfilling the criteria for the certification of their fisheries with the help of cooperation between several nations in their areas or with the help of joint ventures with fishing enterprises from the industrial nations.Along with that, eco-labeling could also provide an opportunity with respect to the innovative producers in order to provide benefits from the application of production methods that are more environmentally friendly.

4.8. IMPORTANCE OF ECO-LABEL

The structure or process of certification in several numbers of countries all across the world, the organization has taken on the very important job of the ECO process. The main objective of the certification is to direct them in five major regions.

These major areas are, system certification, laboratory services, product, and service certification, calibration services and personnel certification, and laboratory services. There is a brief information on these rubrics that are mentioned below:

4.8.1. System certification

ISO 9001 Quality Management System, ISO 14001 Environmental Management Systems, ISO 22000 Food Safety Management System, ISO 27001 Information Security Management System and ISO 13485 for Medical Devices Quality Management Systems, as well as several numbers of management systems for the organizations, certification services checked-in, in accordance with the standard principles of ISO 17021.

This standard develops the settings for the certification for institutions that offers the certification service to the system. Eco-label refers to this standard when authorizing institutions that offer certification services to the system.

4.8.2. Certification of products and services

Eco-label is a certified organization that wants to document products as well as services in accordance with the ISO 17065 standard. Since this service was offered only with the help of Turkish Standard Institute, it is still called TSE certificate, in the present interval of time.

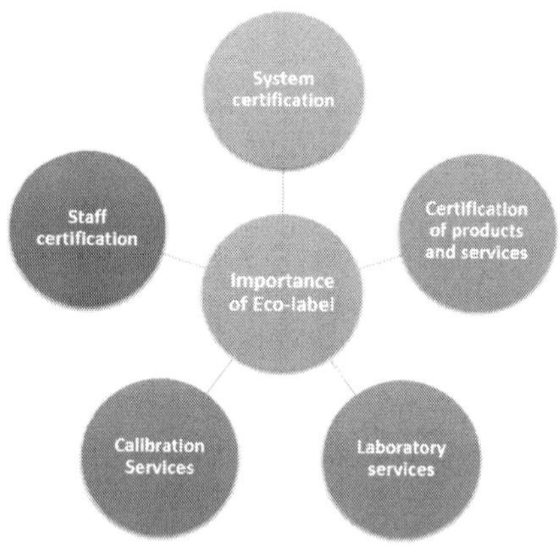

Figure 4.4. *Importance of eco-labeling.*

4.8.3. Laboratory services

Eco-label is a certification laboratory that will conduct several numbers of different studies in the context of measuring, testing, analysing, and evaluating and organizing the reports with respect to the ISO 15189. These certification services make sure that the measurement reliability, measurement traceability and measurement uniformity, and measurement infrastructure availability.

4.8.4. Calibration services

ECO Label accredits laboratories for calibration in accordance with ISO 17025 standards.

4.8.5. Staff certification

Institutions that offer the services confirming the qualifications of personnel that must be delivered to the company within a particular subject and confirming their capability are certified with respect to the ISO 17024.

Some organizations eco-label offers accreditation services to certification and the conformity assessment agencies and the laboratories that will functions in several numbers of the different sectors with a strong technological infrastructure, along with the trained as well as experienced specialists.

4.9. BENEFITS OF ECO-LABELING

Eco-labeling has several numbers of major benefits:

Figure 4.5.*Benefits of eco-labeling.*

4.9.1. Informing consumer choice

Eco-labeling is an efficient process in order to provide the needed information to the customers about the environmental influences of the specific kinds of products, and the choices they can make.

It empowers the people to differentiate among the products that are dangerous to the environment and those more compatible with the environmental objectives.

An eco-label makes the customer more ware about the benefits and the advantages of certain types of products, for instance, recycled paper or toxic-free cleaning agents. It also encourages the effectiveness of energy, waste minimization and the product stewardship.

4.9.2. Promoting economic efficiency

Eco-labeling is typically cheaper as compared to the regulatory controls. With the help of empowering the consumers as well as manufacturers in order to make the environmentally supportive decisions, the requirement of regulation is kept to a minimum. This is advantageous to both government as well as industry.

4.9.3. Stimulating market development

When customers choose eco-labeled products, they have a direct influence on the supply and demand in the marketplace. This is a signal which guides the market in the direction towards a greater environmental awareness.

4.9.4. Encouraging continuous improvement

A dynamic market for eco-labeled products motivates or empowers a corporate commitment with respect to the non-stop environmental enhancement. Customers can predict to see the environmental influences of the products that are decreasing over the interval of time.

4.9.5. Promoting certification

An environmental certification program is a closure of the approval which depicts that a product fulfils the requirements of the certain standards of the eco-label. It also offers the customers with visible proof of the desirability of the product from a point of view of environment.In this way, certification has an educational role for the customers, and also motivates the competition between several numbers of manufacturers. Since certified products have a prominent logo in order to help or to inform the choices of the customers, the product stands out more easily on the shelves of the stores.

And they are also having the desire that the logo can influence the manufacturers to re-engineer the products, in this way, they are less harmful to the environment.

4.9.6. Assisting in monitoring

In addition to this, another advantage of an official program od eco-labeling is that the environmental claims can be monitored or observed much easily.

Both competitors as well as the customers are in better conditions in order to judge the validity of a claim, and will have a profit, to do so, they should claim appear dubious.

4.10. FUTURE TRENDS IN ECO-LABELING

The requirement of actually apply a label, relies on the nature of the relationship between the producer or manufacturer or customer (either an individual consumer or a commercial buyer). The cases of the IFOAM (International Federation of Organic Agriculture Movements) and the FSC (Forest Stewardship Council) are instructive.

In the case of organic agriculture, where the customers themselves play a role of individual consumer, a mass market is being supplied with the help of an entity with which the vast majority of the consumers have very little direct relationship.

The consumer is also unlikely to have the ability to undertake the tests in order to make sure that the product is in fact organic, and perhaps has very minimal potential to sanction the producer if it is found that the producer is dishonest. In this way, a physical label is required in order to overcome a lack of trust between the producer and the consumer.

In the case of FSC, the association is between the producer and a comparatively restrained number of buyers. It is quite likely that the producer and the buyer have maintained a relationship over several numbers of years, and so understand and the trust each other.

It is true that the purchaser of the CFPs (Certified Forest Products) does not have much potential in order to find out whether a product is actually from a sustainably managed forest, but the purchaser has a meaningful potential in order to sanction a dishonest producer with the help of contradicting them future supply contracts. Given the market power of retailers such as IKEA, Home Depot and B&Q, this is not something most producers would be slightly harmful.

The lesson here is that the physical application of a label is not always necessary to the efficiency of an initiative which are based on the market in order to promote a sustainable production.

Labels are likely to be very significant where initiatives encourage the sustainable consumption, but only if the customer is individual consumer.

Sustainable procurement policy is a very significant driver for change. It is not clear that all relationships between the producers and public or private buyers are required to be simplified with the help of application of a physical label fixed to the product. This is extremely vital with respect to the two other trends.

First of all, there has been a slow shift in the focus of the environmental needs from the physical aspects of a single product or product line (e.g., Blue angel) to the management of an entire production facility (e.g., the FSC and MSC, Max Havelaar, the IFOAM) and in the form of corporate codes of conduct and the policy statements, to the overall activities and the influence of the complete companies (e.g., the Global Compact).

No longer are organizations concentrating exclusively on producing a few green product lines. Their complete corporate policy considers environmental as well as social responsibility. More and more consideration to supply chain responsibility has accelerated this evolution.

Second, consumers are more and more focused on a complicated range of environmental as well as social process. They acknowledge that the sustainability (or responsibility) necessitates a fine balance between these kinds of concerns, something that they might not have the time or potential to comprehend.

In the end, the public is turning to what they perceive to be responsible organizations that can do the research and recommend responsible products as well as services. With respect to the several numbers of aspects, this is the role played by the FSC. It is the association with the NGO movement and WWF, more specifically, has made it an unimpeachable source of good advice in the eyes of public.

Companies adopting the FSC label are, in effect, co-branding their products and advantageous from the positive relations that are linked with the FSC and its supporters. As mentioned, the most common reason given with respect to the support of companies of the FSC is not the environmental efficiency of the standard, but the fact that it has been developed in a multi-stakeholder forum.

One of the most important outcomes of these trends in the direction towards holistic corporate social responsibility, and away from a focus on a product lines alone, it that the commercial policy is becoming a much more important factor in spreading environmental necessity as compared to the formal trade policy at least for some types of eco-labels.

On the other hand, WTO's (World Trade Organization) TBT (Technical Barriers to Trade)and the SPS (Sanitary and Phytosanitary Measures) agreements offer a frame of work in order to cope up with these kinds of issues

When they are combined into the trade policy (never the less, imperfectly), there is no frame of work in order to make sure that the necessity of the supply chain are developed, adopted, and executed in a way that does not produce unnecessary blockades with respect to the market access.

This is something that deserves a great deal of attention and could well represent some of the concerns in the context of environmental requirements and market access expressed with the help of emerging nations.

With respect to the environmental standards, whether technical regulations are being force with the help of WTO members or supply chain requirements are forced by the buyers, necessitates the identical basic institutional infrastructure as well as technical assistance.

4.11. CONCLUSION

The certification and the eco-labeling of the product can be a very significant tool in order to achieve the sustainable fisheries and healthy aquatic ecosystems.

They can accompaniment and support traditional regulatory measures in order to achieve conservation and management outcomes.

Eco-labeling of the fish and the fishery products has the ability to develop a market profits in order to regulate or handle the fisheries and aquaculture farms in sustainable manner.

There are several numbers of benefits that can accrue to the world community if this potential is realized that are mentioned below:

1. There will be environmental improvements in the aquatic ecosystems, decreasing the social costs of the reduction in the global biodiversity.
2. Consumers will take advantages as they get more information with respect to the product they purchase, and also, having the potential to choose from more products of various environmental qualities, and are able to make informed choices in the context of the purchase of those seafood products. Consumers also take advantages for a long term with the help of regular or frequent availability of their favorite seafood products.
3. Producers of the eco-labeled seafood benefits from being capable to abstract that additional willingness to pay from the consumers that they would not generally be able to do in an undifferentiated market.
4. The industry of fisheries will take advantages as they move from an unsustainable fishery to a sustainable fishery that preserves production and jobs over the long run.

REFERENCES

1. Carolyn, D., (1999). *Eco-Labeling and Sustainable Fisheries*. [eBook] Huntingdon Road Cambridge: IUCN-The World Conservation Union and the Food and Agriculture Organization of the United Nations (FAO). Available at: http://www.fao.org/3/a-ad349e.pdf (accessed on 8 March 2020).
2. Charles, F. M., (2008). *On the Economics of Eco-labeling*. [online] research gate. Available at: https://www.researchgate.net/publication/228822337_On_the_Economics_of_Eco-labeling (accessed on 8 March 2020).
3. Eco-label, (n.d.). *Importance of Eco Label*. [online] Eco-label.com. Available at: https://www.eco-label.com/en/eco-label-institute/importance-of-eco-label (accessed on 8 March 2020).
4. Eco-labeling of seafood products to reduce environmental damage, (2009). [eBook]*The University of the West of England, Bristol: Science of Environment Policy*. Available at: https://ec.europa.eu/environment/integration/research/newsalert/pdf/135na1_en.pdf (accessed on 8 March 2020).
5. International Institute for Sustainable Development, (2013). *Benefits of Eco-Labeling | Eco-Labeling | Markets*. [online] Iisd.org. Available at: https://www.iisd.org/business/markets/eco_label_benefits.aspx (accessed on 8 March 2020).
6. Mason, C., (2013). The Economics of Eco-Labeling: Theory and Empirical Implications. *International Review of Environmental and Resource Economics*, [online] 6(4), pp. 341–372. Available at: https://www.researchgate.net/publication/257569789_The_Economics_of_Eco-Labeling_Theory_and_Empirical_Implications (accessed on 8 March 2020).
7. The Trade and Environmental Effects of Eco-labels: Assessment and Response, (n.d.). [eBook] Available at: https://unep.ch/etb/publications/Eco-labelpap141005f.pdf (accessed on 8 March 2020).
8. Trevor Ward and Bruce Phillips, (2008). *SEAFOOD Eco-labeling*. Ames, Iowa: Blackwell Publishing Ltd: A John Wiley & Sons, Ltd.
9. Ward, T., (n.d.). Measuring the success of seafood eco-labeling. *Seafood Eco-labeling*, [online] pp. 207–246. Available at: https://www.researchgate.net/publication/229808269_Measuring_the_Success_of_Seafood_Eco-labeling (accessed on 8 March 2020).

10. Yenipazarli, A., (2015). The economics of eco-labeling: Standards, costs, and prices. International Journal of Production Economics, [online]170, pp. 275–286. Available at: https://www.sciencedirect.com/science/article/abs/pii/S0925527315003709 (accessed on 8 March 2020).

5

Fishing Activities and Modes of Production

CONTENTS

This chapter is about the type of fishing gears, trap, and pots used to have a required catch amount. In this chapter, there is a deep discussion on the techniques and methods that changed with the development of fishing techniques. As there are different types of traps and gears are used according to the requirement.

With this, the chapter also explains the mode of production that is used to have specific amount of the species.

The requirement that are needed according to the species. This chapter explains all about the fish techniques and the development in the ancient type of fishing methods.

5.1. INTRODUCTION

A good knowledge of fishing gear is important for fishery management. As there is a great divergence in the efficiency of the different forms of the fishing gears, their ability towards certain conditions, their attraction towards the specific job function.

Earlier there were different techniques for fishing. Traditionally fishing was considered as the art that have been developed over years.

The development helps in the method that can specify the desired fish species and the target size.

The fishing techniques used for different areas is being decided by the test that have helped in choosing suitable choice for the certain area.

The development in the method of fishing have developed specific techniques according to the area.

In chapter, some traditional fishing gears and fishing methods are being discussed that are being occupied by different people around the world according to their needs.

This chapter also discuss about the variety of pots and traps that are used nowadays in different areas of the world to have a specific required catch. As now, the types of traps, pots, and gears are being developed according to the fish species, fish size, and the amount.

This chapter also explains the mode of production of fish species which includes different techniques that are used to reach the target amount of specific fish species.

5.2. CLASSIFICATION TYPES OF GEARS AND FISHING METHODS

Fishing gears are the equipment's that are used by the anglers at the time of fishing. Fishing methods are those methods that are used to have a successful catch. These gears and fishing methods are described below.

5.2.1. Passive gears

Passive gears are commonly known as the passive gears. These types of gears do not need to be dragged, pulled or towed to catch the fishes. Hooks, lines, traps, wires, gill nets are the other effective elements that are used for effective fish catch.

The number of fish that are being catch are recovered by simply removing the gear from the water after certain period of time. At this time, no energy is expended on towing, pulling, dragging of gear.

5.2.2. Hooks and lines

Hook and lines are the simplest gears that are used for fishing. For this type of gear, line, and baited hooks are required. These hooks are present in different shape, size, thickness of wire, type of end and the type of point of the shank.

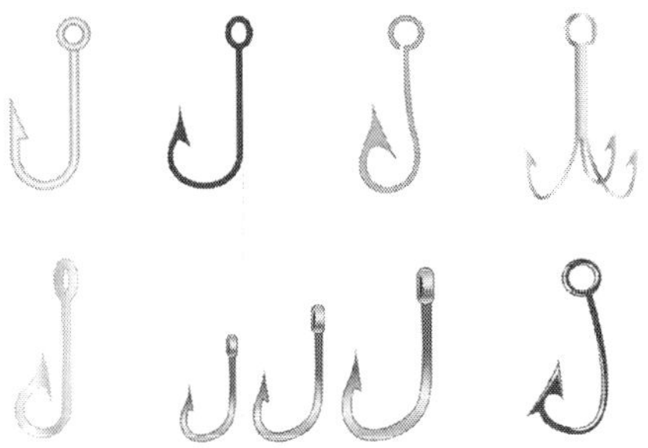

Figure 5.1. Different types of hooks used for fishing.

Source: Image by Vecteezy.

The lines are casted at the place where there is possibility of getting sufficient fish count, there fish swallows the bait and then these fishes are being pulled.

In some countries the line is twisted around the stone and then thrown from the bank or shore into the water, for example in Nigeria this technique is followed mostly (Read et al., 1967). In this case, the stone works as a sinker.

Hooks and lines fishing are considered as less expensive and easy method for fishing. In this case, the catch is alive and also are of high quality. It totally depends on the type of hooks and the size of bait which decides which fish gets captured.

According to the fish variety, a wide variety of sizes of hooks and bait are used. These selections allow to have selective fishing. In fact, this type of fishing includes lots of labor efforts. This method allows only limited number of fish catch per line. Fish are also selected by the type of bait provided.

5.2.3. Setlines

Set line are used to increase the number of lines. These lines can be installed without the need of constant presence of the labor or fisher. Fishing can be conducted in a swallow water or on the beaches. In case of ocean, set lines may get suspended from the surface.

Such lines are needed to be checked regularly as predators can devour the fish that is being caught. This devour of fish can takes place if the lines are not properly and timely checked.

5.2.4. Long lines

Long lines are the version of older hooks and lines. In long lines, main line or ground lines are placed horizontally and is carried by many other small and short vertical gangling where each of them is baited by hook. In the process of line, one end is being anchored to the river or ocean lake bottom and then the position is being marked by the buoy.

The liners then sail along a straight course, which is running out to the full length of the line along bottom.

Then the other end is market by the second buoy. As the line is laid, the hooks are baited. A long line may follow nearly a mile in a length. Each of the lines are often left on the river, lake, sea bottom for few hours before hauling.

When the lines are bought back, the fishes are being removed from the hooks and the lines are then coiled carefully so as to use them again. These lines can be used at the surface also. This is done by suspending in the water column a or fixing it near the bottom.

Fishermen of japan and Italy are said to use surface long line in order to capture tuna, shark, and some bill fishes (Ben-Yami, 1980). Basically, subsurface, and bottom set long lines are used to capture some species of fish like snapper drum bream, halibut, haddock, cod, grouper, hake, and flatfish.

Sometimes, vertical fish stock is used as an alternative of bottom set long lines. This has a rigid branch that allows multiple hooks at a time without snagging. Fishers uses local materials to fabricate the gears.

Long line fishing method offers many advantages. Which involves low capital and investment of energy and labor as intensive operations.

By the hook size, position of the hook in the water, bait type and size help in deciding or choosing the variety of species and size of fish. Long line fishing is commonly practiced by the canoe fishers. If the long line fishery is beneficial in terms of catch, but this technique often considered as laborious.

This is because each hook is need to bait individually and then the captured fishes are being recovered from the hook. The un-baited snagging and entangling long lines are often laid in the deeper portion of the channels. This is done to capture big fishes that are accustomed to rest at the bottom (Welcomme, 1985).

Also, it may be difficult to store long lines and the catch on a smaller vessel. If in case, these lines become caught or entangled on the sea bed, then it will become difficult to recover as this may take many hours. This can also lead the line and the catch to be lost, but this can be solved by using expensive mechanized equipment's.

> The modernization technique in a long line fishery basically involves the mechanization of hauling. This will be effective only when these hydraulic or electrical devices offers better control, variable power and lower maintenance.

Whereas, when the simple mechanical hauling techniques are used for fishing, then in that case they increase the range and depth of long line fishing.

Manually operated reels that are used are constructed of local material. The manually operated reels are used to reduce the effort needed to handle the gear. In some of the boats, a vehicle wheel is used by fitting it with the handle and mounted on the boat. This mounting help in creating a simple roller which can facilitate the handling lines.

5.2.5. Maze gear

Maze gears consist of equipment that are used for leading fish into the situation or in a condition from which they cannot escape easily or from which the avenue of escape is not readily superficial.

Maze gears contains variety of fishing pots, fyke nets, traps, pounds nets and tidal weirs with other modifications.

5.2.6. Pot gears and traps

Traps are the portable traps in which fish enters basically from the small opening. This trapping can take place with or without the presence of bait. Pots are basically used to capture slow moving creatures that may exist for the most part or can be found on the above of the river, lake or sea bottom. The pot gears that are used are especially small in size.

The pot gears are kept small so that large number of catches can be easily loaded on the deck of small boats. They are also kept small so that they can be easily hauled on the ship. These pot gears are either kept half-round, rectangular, conical, cylindrical, etc.

These shapes are made up of different materials such as strips of raffia palm, broom sticks, chicken wire gauze, bamboo, cane sticks among others (Reed et al., 1967; Ectcheri and Lebo, 1982; Welcomme, 1985).

Traps consist of one or two chambers that are funnel shaped and non-returnable valves. With the presence of funnel and chambers, traps also consist of door that is use at the time of recovery of captured species. At the time of operation, in order to weight down the trap some objects are used that are heavy.

These objects can be bricks, concrete, flat stones or sinkers and it can be buoyed line, these objects are attached at the corner of the chamber end.

The baits are then kept on the hook or it can be kept in a mesh bag that is attached to the center. This is done in order to hangs the baits in the chamber. In some of the systems like Anambra and cross river systems, conical traps are used with the different baits like palm fruits, dead toads, cassava, meat scrapes and many other baits like that. These baits are used to lure the fishes.

The buoy line contains a small float that is attached about 2 meters from the pot. This is done to keep the line from becoming cut or to avoid the entangle on the rocky bottom. The surface end of the line is then attached to a small buoy.

There is some disadvantage from the pot gears and that is, it makes the handling difficult and manual hauling also becomes arduous. The major disadvantage of this technique is the high loss of pot gears.

This loss take place due to storm damage, degradation of the materials, due to thief's and due to inability of locating the gears after it gets deployed with the buoy lines. This loss of trap can lead to the depleting of fish resources.

In order to provide attractive shelter and shade for some fishes, mangrove roots, branches, and many submerged vegetation's are used to cover the traps. The use of 'curiosity trap' which have the curved surface are used throughout the Caribbean which helps in attracting the snappers, groupers, and other dwelling fish species who lives at the bottom of river or ocean.

There are various materials and designs that are used for the pot gear and traps that are available throughout the world. Conical traps are said to be made with broom sticks, slashes of raffia palm and cane sticks with other materials that are used in the areas of Niger, Anambra, and cross river basins (Reed et al., 1967).

In Caribbean Island fish traps are found to in-corporate the external funnels that leads in the conical shape of interior funnel.

Interior funnel is being kept in conical shape so as to prevent the escape. Whereas in countries like Hawaii, to catch the deep-water shrimps, trash cans are used. At the bottom of the cans, a cut is made of a 10 gallon and is then replace with the ¾ inch nylon fish net funnel. This way they make the entry of fish in the can.

Then a part of the lid is also cut out from the can and replaced with the ½ inch wire mesh (Bostid, 1988). In order to trap the octopus, earthenware pots are used with handle or sometime without handles. In the country like japan, to catch the octopus PVC (Polyvinyl Chloride) tubes are used as a pot instead of earthen pots.

But sometimes old tires can be also used as a pot that works equally effective. The tires are cut in the rims and then these rims are laced together with the wire and after then a circular wooden disk is being nailed to one end. Even in some places like Venezuela (Bostid, 1988) tire pot gears are extensively used.

Whereas, metal traps are used in the USA. But in case of metal traps, corrosion is the main and serious problem. To avoid the corrosion, use of marine mesh can be the solution for this problem.

This marine meshis used with combination of steel wires which help to strength the wires and they also become durable due to presence of plastic. These types of traps are often expensive and very bulky (AGFSC, 1986).

5.2.7. Weirs, trap net, and fyke nets

Fyke nets are essentially made shallow water gear. This shallow water gear made the fyke difficult to set in the deep waters. Fyke nests are extensively used in the rivers and in the shallow coastal water for fishing.

Weirs and trap nets are the most adapted technique to catch the fish. They are used in fair current conditions. They are sometimes used without any wings or leaders (Von Brandt, 1984).

The simplest form of fyke that is used commonly is a long net bag with a rectangular opening, these such side are lashed to a stake. If there is a current flowing into the opening, then such nets are considered to catch number of different varieties of species. Fyke nets usually contain short vertical net wings that are set indirectly on the either side of the mouth of the bag.

As the fishes are moving with the current strikes, there wings get bend toward the mouth of the net. The fyke nets are used at shore for fishing. They are especially used to collect the flounder and bottom fishes.

In this type of situation, long wings can be used as the current is comparatively slow. The wing is landed on the bottom because of their heavy weight. Wings also consist of cork and glass that floats on the top.

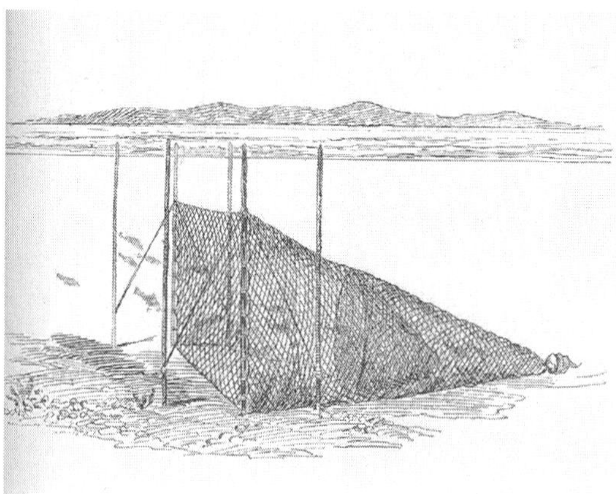

Figure 5.2. Fyke net used for fishing.

Source: Image by Wikimedia Commons.

The nets are set with anchor instead of using stakes. Fyke nets are considered to contain number of advantages that favors the use of these nets in the developing countries. These nets can be constructed by using local materials. This process has little or no energy that can be used for fishing. At the time of season, the number of catches become huge.

These kinds of traps are used only to capture large number without a costly investment in vessels. In this kind of traps, the fishes are captured live and in an excellent condition. It is true that to operate these traps its easy, but initially these traps require a great amount of labors during installation. These are basically used by the people who catches fish physically in a suitable area and to local and seasonal behavior.There are many types of traps designs that are used. In South Asia, large bamboo weirs are used. These bamboo stakes traps consist of long fences with the split bamboo that are secured at the interval to large wooden poles that can be used to drive into the sea bottom.

Long wings help to guide the fish into the heart shaped center, this help to harvest the fish in the net. it is usually set with the open end towards the shoreline. This method is very effective for catching the fishes those are migrating along the shorelines (BOSTID, 1988). Like in Caribbean, in the shallow shelf areas a simple fyke net is used instead.

The length of the wings is decided upon the characteristic of the region, but these are generally with maximum of 4 meters in depth. One of the wings runs from the shores to the corral opening that helps in intercepting the migration path of the fishes.The other wing is placed semi-circular to prevent their escape and in order to direct then towards the corral. From the corral, there is a narrow opening which leads into the smaller holding of pool or crib with the net floor that makes the fish recovery easier.

In Japan, floating fyke net trap is used especially. In this country, the floating net cage is anchored over the shallow shelf. In this case, one wing can be extended up to 100 meters towards the coast.

Whereas, other wing is sent in the deeper water. In this case also, schools of fishes are being guided by the wings into the cage. It totally depends on the placement of the floating trap which can be used for catching demersal and pelagic fish species easily.

5.2.8. Fish fence

Fish fence are installed to intercept the fishes who are migrating upstream. When these fences were made, many fishes who were migrating tries to slide and tries to pass through the barrier.By doing this, many of fishes gets injured while passing through the fence. This process involves thousand of labors, those who spent hours in attempting to net the standard amount of fishes and to move them past the rockslide debris.

Fences are applied to those areas where the level of water lies from low to medium. When the water flow is high, the design allows the water and debris to go over the top of fence.

It totally depends on the design of the which enables the fence to remain in the river for year around with the minimum damage, when the level of water is high.

Figure 5.3. Image of a fish fence applied to the area containing low water level.

Source: Image by Pixabay.

When the fence is established, this allows fish to bypass the fence by entering into the large live box which is with a one-way door system. After then staff can net the fish from the trap to a waiting trailer. Afterwards, the box is then moved up beyond the rockslide with minimum stress.

5.2.9. Fish shelter

Fish shelter are made up of triangular plot of branches that are staked firmly in the river bed and with the apex of the triangle downstream. Each shelter is about 10 meters long and across the base it can be 4 meters. These shelters mostly attract small fishes.Which in turn bribes the larger predators. The fish shelters with fishes, commonly collects the food scraps, kitchen waste which are placed amongst the branches. The fish fence is set around the fish shelter, this way they encircle the fish. Fishers then push inside and remove the branches.

They reduce the area of enclosed space which gradually shifts the fence until the fish become imprisoned in a small space where they are captured with a clap net. The fish shelter is then rebuilt.

5.2.10. Fish trap

Fishing is one of oldest ways by which people feed their families. The technique of gathering of shellfish and spearing fishes by hand, primitive method is also very old and famous used technique of fishing.

In early times, the flowing water which causes high tidal in the rivers and lake, this cause in the change of levels.

When these changes occur, fishermen use to trap fishes behind the rudimentary barriers. These are often made from the sticks and stones.

The idea of fish trap was bought by the fishermen when they found that fish catching can be improved by forcing the fishes into these barriers. Afterwards, they found that catches from these barriers decreased over time as fishes gets familiar to this technique.

So, they understand that they have to shift the trap to the freshwater areas. The idea of moving the trap from still water to freshwater can help in catching more amount of fishes.

The task of moving or constructing new trap involves lots of hard work. That can include from moving the stones from the old trap or finding the new stones for the new trap.

In primitive fishing, fishermen try to make barriers from lighter material which can be more readily available.

For example, using of tree branches, bushes, and vines that are easily available. This helps the fishers to invent more light, movable traps that are made from the bushes and nets that are made up of vines.

This made them to carry their net and traps along with them to the new places. There innovations can make them to try bigger and even more complicated corral type of fish traps in lakes, rivers etc.

Fishermen then found that:

- Fishes are being caught in traps, as the tide fell. These tides forced the fishes into them by the current or could be chased into them by the fisher;
- Fishes enter the trap for protection or simply follows other fish seeking shelter;
- The fishes get attracted with the Objects in the traps such as white stones; and
- Bits of fish or meat can also attract more fish.

From such beginning, the modern traps and pots are being developed.

As the requirement of fish increased, fish become an important part of their food. The requirement helped in making many types of traps and pots. Some of the variety of traps and pots are being discussed below.

5.3. TRAPS

Traps are the simple passive gear used for fishes that allows fishes to enter inside it and then makes it hard for them to escape. This is simply achieved by:

- It is done by simply putting chambers in the trap or pot. These chambers get closed, when the fish gets inside those chambers; and
- To make the funnel in such a way that it makes difficult for the fishes to escape.

Small traps are generally covered except the entrance is made open. On the other side, the larger traps that are extended above the water level are often left open from the top.

5.4. TYPES OF TRAPS AND POTS

In general, traps are the larger structure that are fixed to the shore. Whereas, pots are the smaller, movable traps that are enclosed baskets or boxes that are set from a boat or it be used by hands also.

Traps are simply produced by Von Brandt in the year 1959 for FOA. General type of traps and pots include:

1. Traps that are used for creating barrier for the fish movement, includes walls, fences, gratings, fyke nets and watched chambers

that gets closed by the fishes when they enter the trap.

2. Hidden traps are also used. These traps include bush traps, octopus pots that are used in the area where there is low water flow.
3. At some places tabular traps are also used. These types of traps are basically narrow funnels or hoses that helps the fishes from getting out backwards. Example, eel tubes.
4. There are some traps like gravity traps, bent-rod traps, torsion traps, box traps and snares. These types of traps are closed by fishes mechanically.
5. Baskets, include pots made of wood, wire or plastic, conical, and drum-like traps made of netting with hoops and frames (e.g., drum nets) and box-like traps made with strong frames. These baskets are enclosed traps and pots usually with a structure to make escape difficult.
6. For fishing, sometimes large open traps or corrals are also used. This mechanism is used to stop the fish to escape from the trap. These are fixed on sticks or anchors or sometime they are set or left floating.
7. Traps are sometime set out of the water. This is done to catch the fish such as flying fish that usually jumps off the top of the waves and glide over the surface when in danger.

These traps can be of box-shaped, rafts, boats or nets. Scoop nets are sometimes used for making fish jump. Pitfall traps can be used for marine animals that migrate over land, such as coconut crabs.

5.4.1. Entangled net and gill net

Gill nets are basically rectangular walls made up of net which is kept straight by means of floating and sinkers. They are positioned in the swimming layer, in the way to target the fish. Fish is being caught by holding them in the mesh by galling.

Whereas, drift nets are operated in the surface layer and then it is drifted with the current, that can be separately or can be with the boat to which it is tethered. Set gill nets or anchored gill net are fixed at the bottom by using anchors or ballast.

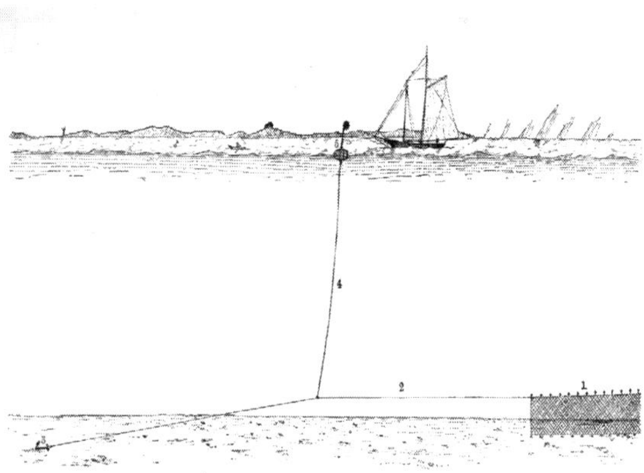

Figure 5.4. *Image of a fishing gill net.*

Source: Image by Wikimedia Commons.

Fixed gill nets are basically operated in the shallow coastal waters. They are being fixed by the means of stakes, then the catch is collected in the presence of low tides. Encircling gill nets are operated in the surface layer in coastal areas.

Fishes are first encircled, then the noise and other vibrations are used which help to drive the fishes towards the net. This way they are either gilled or encircled.

The gill net is based on the structure. There is single gill net which contains single wall of net that is supported by the floats and sinkers. Triple-walled gill nets also called as trammel gill nets are generally operated as bottom set.

This type of gill net has two outer wall which are larger mesh size and are loosely hung inner wall of small mesh size.

When the fish approaches through the large mesh on the outer wall get intercepted by the by the inner wall. this forms a pouch after passing through the large mesh on the outer wall on the opposite side. This helps in holding the fish properly and securely. In combination, the lower part is fabricated as a trammel and the upper part is simply a gill net.

Entangling nets are loosely hung in single or multi netting. These nets are held vertically in water by floats and sinkers. These nets are used to catch fish by entangling rather than enmeshing. These nets are attached end to end in order to form large fleets.

5.4.2. Set gill net

Set gill nets are stationary gill nets those are fixed in a position of strategic place. These nets are used suspect the fish when they pass through. Set gill nets are positioned by using anchor or stakes. These nets are useful for low water flow and demersal fish.

Gill nets visuality during day time by fishes makes them less effective. Thus, the fishing is usually done at the night (von Brandt, 1984). Sometimes, the presence of turbidity also renders the effectivity in daylight.

If the water contains many of the phosphorescent organisms, the gill nets might not be able to catch the fish effectively during nights but can be more effective during twilight. Mostly, the catch at the night time may get dead or can be eaten by the predators before the nets are hauled.Set gill net are the most effective gear used that can be helpful in catching the fish throughout the year with different type of species. The factors like length, width, and mesh size of the net help in deciding the amount of catch by set gill net.

5.4.3. Trammel net

Trammel nets are similar to the gill net. The one difference is that this is made up of three layers of net. In some countries these nets are set on or close to the sea bed that help in catching or targeting the demersal fishes.But sometime in case of overseas fisheries, these nets are being set in the mid of the water or just below the surface in order to target the pelagic fishes.

Figure 5.5. Image of a trammel net set in the sea bed.

Source: Image by Wikimedia Commons.

5.4.4. Lift net

Lift nets is the major fishing gears types that can be used in both the municipal and commercial fishing sectors. The basic commercial lift nets are bag net and round haul seine that are racked fourth and ninth among the commercial gear that contributes only 8.5% in 1995 commercial fishery production.

The municipal lift net is basically composed of the beg net and crab net lift nets that contribute 1.88% in the year 1995 municipal fishery production.

5.4.5. Crab lift nets

The gear consists of a sheet of net that is supported by the bamboo frames. This is done to capture the crab by entangling as the net gets lifted to the sea surface. The frame is made in a square pattern which is almost 40–60 cm along the sides.

But the height of the frame can vary in height ranging from 7–20 cm and measures 70 cm by keeping diameter of 1 cm. The net is made up of either nylon monofilament or multifilament 210d/4 with the mesh size of 30–76 mm. The retrieving line is made with Styrofoam or bamboo floats which is tied at the center of the frame. At the tip of the bamboo frames, stones or lead weights are tied.

During operation the bait is tied at the center of the frame. Lift nets can be operated or can be set without bait. The fishermen can operate the gear in shallow water. This can be simply done from a rowing banca or by simply walking along the shore.

The lift nets are being observed very often to check the catch amount. This catch basically contains the mangrove crabs and blue swimming crabs. Crab fishing can be done all year around and can be operated during day and night time both.

5.4.6. Shrimp lift net

Shrimp lift net look like the structure of the crab lift net. But these shrimp nets are having longer frames and bigger nets. The bamboo frames are long by 1.5 cm which measure 2.6 meters with diameter with 1.1 meters of height.

The net is made up of nylon mono with 0.7 mm or it can be made from nylon multifilament 210d/4 with 10 mm of mesh size. In structure, it is basically in square shape which is 75 cm from all the sides. Along with bamboo frames that size 60 cm above the net, stones are also there weighing about 100 to 200 grams.

The gear is provided with the boiled fish which is used as a bait at the center of the frames. These nets are set along the river during night time with one fisherman setting 10 units. Sometimes, the caught shrimps and goby are used as a bait for the bottom set long lines.

Figure 5.6. Boxes or traps that are used to catch the crabs.

Source: Image by Needpix.com.

5.4.7. Fish lift nets

Fish lift net the most advanced and diverse type of lift net. It totally depends on the type of boat and fishing area which decides the type of particular technique and methods of operations. The simplest form that is used is a small outrigger, motorize banca.

It totally depends on the boat size, outriggers, and length of the extension broom which helps in deciding the size of the net. The mesh size is decided by the size of the fish that is needed to be targeted.

A unique method that is used for operating small scale bag net of Zambales. The net used is kept 10.8 meters long 9 meters wide and 4.2 meters deep. Fish lift nets are made up of nylon multifilament 210d/4 with 8 mm as mesh size.

The wooden outrigger banca has a length which lies from 8 to 10 meters and powdered by a 10–16 hp gasoline engine. This type of lift net is managed by the two fishermen.

When the boats reach at the fishing ground, the banca is then anchored and starts lighting which uses two kerosene lamps of 500 watts each. First the sufficient amount of the fish school is attracted and then the lamps are being transferred to a bamboo floats with the 20 m retrieving line. This retrieving line is allowed to drift for some distance from the banca.

Then the bamboo poles are being prepared by the fishermen which 7 meters long with the base containing diameter of 6–8 cm and diameter of 3–4 cm of the end portion. The pull ropes are tied at the end of the bamboo poles which are passed through the pulleys.The net is then lowered below the banca. Simultaneously, the lamps are retrieved to their original position. In order to concentrate the fish school, a conical shaped covering is placed around the lamp.When the signal is being given to the fishermen, they start hauling the net until the catch is bought on board. This process can be done 3 to 5 times per night. Using this net, the species that can be caught are anchovy, mackerel scad, small caesio and sardine.

5.4.8. Stationary lift nets

These types of lift nets are being set in the shallow waters. Where the depth of the gear is totally depended on the length of the bamboo or palm stakes which serves as the posts.

The frameworks are said to consist of two or more rows of bamboo that are set parallel to each other. They consist of four equal sides that are permanently stacked on the seabed.The two rows of bamboo are being supported by the braces and catwalks. These rows are one meter above the highest tide where fishermen lift the net. The posts are provided with the pulleys and rings. Through these rings and pulleys the pull ropes passes through.In some structure, for hauling operation some manual winch is being installed where a platform is constructed. Other areas have corner posts which is provided with the vertical rope that weight 50 kg at the end where during hauling, the rings of the pull ropes passes through.

The length, width, and depth of the bamboo structure or frameworks decides the size of the net. In general, a stationary lift net has a length of 10–15 meters and contains a width of 10–14 meters with the 5–10 meters depth. The material that can be used for the net can be PE 380/6–9 with mesh size of 0.2 mm. minnow net 400d/6 with 4 mm mesh size or nylon multifilament 210d/9 with mesh size of 16 mm.

5.4.9. Round haul seine

Round haul seine gear is a type of lift net where operationally it consists of 3 non-motorized light boats, two flatboats and one out trigger motorized banca. When the boats go to the fishing ground, the out trigger banca tows the two flatboats and 3 light boats.The light boat gets anchored at 300–500 meters in distance from each other. While the two flat boats wait in the area. At the time when sufficient fish school is being detected, the light boat signals the flat boat to prepare for the operation.The float boats then proceed near the light boats with catch school fish. After then they drop the anchor taking into the account of current direction.

The flat boats transfer one side of the net to the other flat boat. This way they separate the sideways by spreading the net.The flat boats are 13.7 meters long and are powered by 16 hp of engine. The light boats are considered as non-motorized dugout canoes with the length of 4.9 meters. There are at least 2 or 3 pressurized kerosene lamps that represent on each light boat. 14 meters long service banca powered by an 80 hp of fuzo power. The main net is said to be made of knotless nylon 210/6 with the mesh size of 8 mm. Selvage uses polyethylene 400d/18 with the mesh size of 33.7 mm. In length, float line measures up to 73.2 meters in length that is made up of polyethylene rope f 10 with 40 Styrofoam floats.The lead lines contain length of 36.6 meters and are made of diameter PE from PE f 10 mm. net on both sides are provided with 6 pull ropes. But out of 6 only 3 ropes are provided varying weights that can be 15 kg, 30 kg, and 40 kg.

5.5. MODE OF PRODUCTION

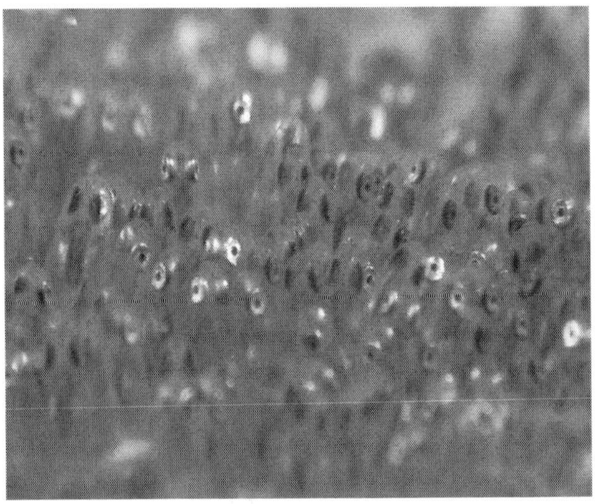

Figure 5.7. Fish breeding in a pond.

Source: Image by Big Al's Pets.

5.5.1. Fish breeding

Fish breeding is similar to the fish propagation which means multiplication. Breeding includes all the various techniques and methods which can be widely used to maintain the fish stock.

This technique is used widely in fish production when the demand of fish gets increased and for the fish culture. In tropical regions, many of the fish species are breed only once in a year. A continuous demand for fish breeding makes fish propagation sine-qua-non. Basically, there are three types of fish breeding/ propagation:

- Natural propagation/ breeding technique;
- Naturally induced breeding/pond or enclosure propagation; and
- Artificial propagation through hypophysation.

5.5.2. Natural breeding technique

Natural breeding technique is where fishes can breed in their natural environment. Then their eggs, larvae, fry, and fingerling are being collected for pond stocking. The fry and fingerling of Cichlids, Chrysicthys as well as mullets are available in Nigerian waters in a large quantity all the year around.

The collection of the fish seeds is carried out with no physical damage to the seed. This entails the elimination of the gill-nets and hooks are used as gears.

Gears that are used for collection includes the hand net, lift net, drag nets, cash nets. Where Light is used to attract the fish at the night. Sometime GNC are used to attract the fish to avoid the mudding during seining. There are some problems that arises during natural breeding, those are mentioned below:

- For many species, the seed may not be available as and when needed or required. Hence, there can be shortage of fingerling for stocking leading to low fish production.
- There are some species of fishes, e.g., Dragonfly larvae, water bugs. These species may feed on eggs or attack the fry or compete for the fish food, which may also be collected with the fish seed.

Fish parasites, e.g., leeches are collected with fish seed from the wild and introduced inadvertently into the rearing ponds.

- Difficulty of accurate identification of the fry/fingerling stages of certain species. Which results in the desired species being stocked with undesirable stunted species.

- There is high mortality during catch/collection and at transportation.
- This breeding technique may be uneconomical, e.g., cost of going to the wild, pay workers to gain access to the spawning sites through trial and error methods.

However, with all these disadvantages, the fish can breed naturally without much expending is a great advantage.

5.5.3. Ponds or enclosure propagation

Essentially it a naturally induced breeding technique that involves introduction of brood fishes or brooders into specially constructed pond or tank in which the breeding is allowed. This technique is practiced to leave the brooders breed repeatedly.

This technique is not accepted in terms of quality and is anti-selective. As the age and percentage of the fish cannot be determined after certain time. A proper and systematic pond propagation technique that ensures the production of fry of known age is highly recommended.

For this technique, ponds that are used for propagation are basically small in size which ranges around 0.1 ha or 100 m^2 and shallower less than 1 m. Cement or concrete tanks measuring 5x3x1m are used mostly for the propagation.

When tanks are used for propagation, then they are covered with the layer of sand which is done to make their nests.

The eggs gest hatched within 2 to 3 days after external fertilization. The fry moves in school along the edges of the water from where they can easily collect.

5.5.4. Artificial propagation through hypophysation

The hypophysation technique is basically done to induce the spawning in fish by using pituitary gland which can carried out any time of the year and under any environmental condition. This technique ensures that ensure the availability of the fish seeds all the time of the year.This technique can be used for single common crap that has been induced to spawn five times in a year at interval of 60,62,41 and 186 days. Although the crap breeding takes place only once in a year. Artificial propagation was first described in the year 1765. But it was neglected until the year 1842, when this technique was rediscovered.Artificial propagation was first reported in Panyan fish farm and Agodi fish farm. In these farm craps are being propagated successfully. There are other fish farms that have tried the hypophysation using catfish. Example Clarias Gariepinus, Heterobranchus Biodaorsalis.

Artificial propagation can be divided into 5 selection of brooders:

- Maturation of the brooders;
- Stripping i.e., obtaining eggs and sperms (milt) from the brooders;
- Fertilization;
- Incubation of fertilized eggs to ensure that they hatch; and
- Rearing of the larvae/fry up to fingerling stage.

Advantages:

- Fish seed is guaranteed all the year round;
- Fish seed is obtained outside the natural environment of fish;
- It increases the survival rate of the fry; and
- It improves quality by crossing two different species (i.e., hybridization) can be obtained.

Disadvantages:

- The donor fish must be sacrificed in most cases and hence of loss of fish;
- The whole process is laborious and highly technical; and
- Very expensive, as it requires proper housing, constructions of tanks, installation of jars in a close circulatory system.

It should be noted that the artificial propagation method that is naturally induced or through hypophysation, The production of fish seed is carried out in enclosure known as Hatcheries. This may be an indoor or outdoor facility. This type of propagation requires inputs such as brood stock, adequate water supply, and suitable feed. There is sequence of steps that are involved in artificial production of seed.

5.6. MANAGEMENT/MAINTENANCE OF BROOD-STOCKS

The maintenance and management of the brood stock on breeding farms are necessary. The maintenance of the brood stocks is necessary because it permits the selection of healthy, high quality brooders and helps in building up in order to improve the stock.

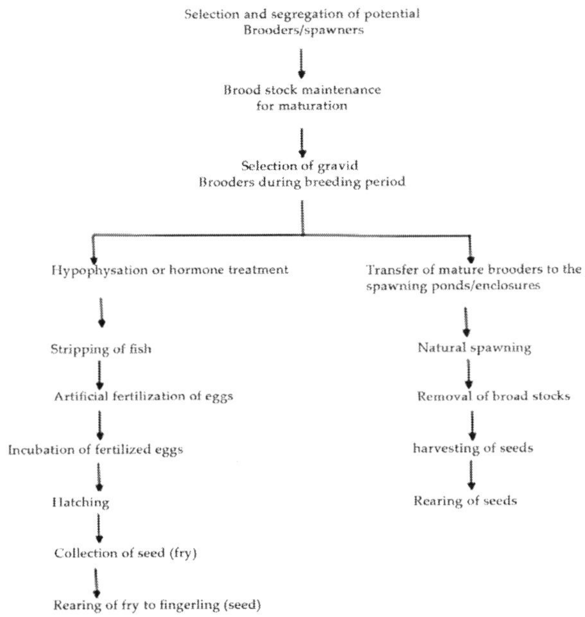

Figure 5.8. Planning for sequence of steps involved in artificial production of seed.

The ponds that are used for keeping brood stocks are first needed to be simulated to the fish preferred habitats which include oxygen content, temperature, pH, size, depth of the water and tranquility, stocking density, quality, and food availability.

The sharp fluctuations in the physico-chemical parameters of the water in the pond must be prevented. Especially temperature and oxygen are the physico-chemical parameters that rare known to inhibit the gonadal development.

Overcrowding in the pond condition must be avoided because it consists of negative stress on the stock. Although there are many cultured species that can tolerate crowded pond situation at some degrees. For example, brooders of catfish that are held at a stocking density of one fish in the pond area of 2 to 5.0 m^2 with depth of 1.0 to 1.5 m.

Frequent disturbance interferes with the normal gonadal development. Therefore, an extra care must be taken to restrict the seeing or netting of the fish to minimum. For example, Claris can withstand being netted out of water twice or thrice in a week. Whereas other fish species should be seined for more than a week.Suitable and adequate quality of food is equally of great importance. It is being known that brooders raised or maintained on adequate natural food or protein-rich artificial feeds that have a higher fecundity with larger eggs and thus resulting the best yield.

Suitability and adequate quality are equal and play a great importance. The brooders are nurtured and maintained using natural food or protein rich feed that are artificially developed to have a higher fertility of eggs that can yield to high and best quality result.

Brooder should be treated and bathed routinely in a 25% NaCl or with 150 ppm formalin solution for about 10 to 30 minutes. Brooders need maintenance because to prevent the outbreak of the diseases in the stock that is used for breeding or can be also subsequently transmitted diseases to the offspring. For preventing the brooders malachite green can also be used.

There are factors like species type, size, and number of eggs per kg per female, the survival from the fertilized egg state to fry stocking phase, demand for the fish from the grow-out unit and many other factors are responsible for deciding the procedure of maintenance or management of brood stock.

The principal of brood stock management help in solving following problems:

1. The number of eggs of the specific species that is required to produce the fry;
2. The average weight of the female Clarias. Calculate the number of female brood female stock to be maintained;
3. Maintain the sex ration of male and female that effects the breeding with the ration 2:3. This helps to calculate how many males would be required; and
4. In order to have proper maintenance, at least thrice the required brood stock must be maintained in a brood stock pond.

To maintain the brooders, it is required to calculate the surface area of the pond. That helps in ensuring the above fry supply.

The larval stage ends when they fill up their air bladder with air and begins swimming like a fish. They also start to eat external food then it becomes fry.

In addition to all the essential that are needed by the larva, the fry also requires external food. The external food that is provided to the fry should be adequate in terms of both qualitative and quantitative.

In case of early fry, they still have a part of yolk left and they can draw on it for sustenance from 1 to 4 days which totally depends on the type of species. In this period, fry learns to find its own food.

The fry requires a more precise and careful nursing that will ensure the proper growth and survival of the fry. Lack of suitable food availability will lead fry to high mortality.

Fingerling are bigger than fry, the size can vary between 5 to 10 cm. At this stage, the management of these development stage is totally based on their fragility and size difference in terms of their habitats, stocking, density, feeding, and control on the infection and diseases. This stage is crucial as they cannot be harvested without recruitment. This stage is called as fish seed or recruits.

Fry are nursed in a small earthen pond. The size may differ from 100 to 200 m^2 for 3–4 weeks in order to attain the fingerling stage. The fingerlings are nursed in a bigger earthen pond. Ponds are usually prepared to have a standing crop of rotifers.

They must be checked properly to avoid or to exclude the cyclopoid, copepods which are the natural enemies of fry. Fry usually have two sources from which they prepare their food during initial stage. Yolk and external food are the two ways through which they fry prepare their food to ensure better survival.

After completion of one month, the young fingerlings are removed from the small nursing pool. They are then shifted to the large ponds. It is important to thin out the stock because that helps in providing enough space, oxygen, and food to ensure better growth. At this stage, enemies must be managed too. There is always change in the enemy with the age of the fish. These can be classified as:

1. Enemies of fry;
2. Enemies of advanced fry; and
3. Enemies of fingerlings.

It is important to identify these enemies on time and treat them adequately. The insect larvae are said to have greater danger at this stage which is followed by hunger if there is a shortage of food.

If there is deficiency of oxygen in the pond then that reason is enough to kill the organism in a highly manured pond. If there are abrupt changes in the temperature and presence of extreme cold can be responsible for eradicating the fry population.

The pond infection with any of the protozoan or bacteria or fungus can be easily diagnosed by some certain indicative signs. Some of the signs are discussed below:

1. Swimming of fry or fingerlings in a large school near the surface;
2. Their gathering below the water inlet;
3. Occurrence of the dark specimen; and
4. Sudden occurrence of the dead fish on the surface of the pond.

It should be kept in mind that fishes those die due to presence to parasitic infection floats on the surface of the water. Whereas, the fish that die due to dietary factors, remains at the bottom of the pond.

5.7. CONCLUSION

There are different types of fishing methods that are used nowadays. Earlier techniques involve fishing at a very small scale at a time. This means that fishermen use to catch the fishes, which involves on netting or basic technique for capturing the catch.

Which does not allow to have a greater number of catch at a time. As the amount was not fixed, there was a possibility to have large number of fishes that can be wasted if not required.

But now, the latest techniques are used which contains different types of nets and gears that help in achieving the target amount of specific fish species. Some of the examples of gears types are passive gears, maze gears, pot gears, etc. There are many other types of gears and methods that are used to fulfill the requirement.

Similarly, there are different types of traps and pots like set gill nets, lift net, portable gill nets, etc., that are used to have the catch in a specific area.

Now, as the requirement for the fish is increasing. There is also important to maintain the availability of the fishes in the area. So, to have main the availability of the species, the different mode of productions is used to grow the variety of fish species in the required amount.

Fish breeding, natural breeding, propagation, and many other techniques are used to develop the species of fishes in a specific area by providing the specific growth requirement.

REFERENCES

1. Agritech.tnau.ac.in. (n.d.). Fisheries: Home. [online] Available at: http://agritech.tnau.ac.in/fishery/fish_fishingtech_passivegears.html (accessed on 8 March 2020).
2. 2005. BASIC FISHING METHODS. [ebook] Available at: <http:// www.vliz.be/imisdocs/publications/265677.pdf> [Accessed 8 March 2020].
3. Eyo, J., (1995). *Fishing Gears and Fishing Methods*. [eBook] Nigeria. Available at: https://www.researchgate.net/publication/233817982_FISHING_GEARS_AND_FISHING_METHODS (accessed on 8 March 2020).
4. Fao.org. (n.d.). *Figure 1*. [online] Available at: http://www.fao.org/3/x2590e/x2590e02.htm (accessed on 8 March 2020).
5. Map.seafdec.org. (n.d.). *Monograph*. [online] Available at: http://map.seafdec.org/Monograph_project/lift_net_4.php (accessed on 8 March 2020).
6. Omoniyi, D., Agbeti, D., & Olaoye, D., (n.d.). *Fish Production and Management*. [eBook] Available at: https://pdfs.semanticscholar.org/3601/ca927487c7e1b8978f5d4b715c754ac9b884.pdf (accessed on 8 March 2020).
7. Seafish, (n.d.). *How to Make the Most of TESS*. [online] Available at: https://seafish.org/gear-database/fishing-gear/ (accessed on 8 March 2020).
8. Seafish, (n.d.). *Trammel Nets*. [online] Available at: https://seafish.org/gear-database/gear/trammel-nets/ (accessed on 8 March 2020).
9. Seymoursalmon.com. (n.d.). Fish Fence -Seymour River Hatchery. [online] Available at: http://seymoursalmon.com/projects/fish-fence/ (accessed on 8 March 2020).

6

Conservation and Management of Marine Fisheries

CONTENTS

The chapter of conservation and the management of the marine fisheries explains the basic significance of the marine nature conservation and management of the marine fisheries. This chapter also mentions the areas that are marine protected for nature conservation.

This chapter is consisting of problems and issues that are faced by the fishery management such as conventional fishery management, fishery paradigm, conventional management tools, and effort controls. This chapter also explains the problems that comes in evaluating the regulatory systems.

This chapter emphasizes regulating fishing reports that is consisting of closed season, conservation, economic performance, and social value. Also provides highlights on the effectiveness of the management of the marine fisheries all across the world.

6.1. INTRODUCTION

All across the world, the conflicts between marine nature conservation and the fisheries interests are common and increasing, and there is every so often a glaring absence of dialogue between stakeholders who are representing these two interests.

There is a requirement for a stronger and the imposed coordination between the fishing and the conservation authorities when developing the marine protected regions for the reasons of conservation.

Researcher proposes that a suitable instrument for these kinds of coordination is a broad marine spatial procedure for planning which is based on the ecosystem, that represents neither nature conservation nor fishery.

Strategic Environmental Assessment (SEA) for planning and programming and Environmental Impact Assessment (EIA) for projects are generally utilized several numbers of tools in order to assess the environmental impact of various human activities, but are not often used in order to evaluate or assess the environmental effects of capture fisheries.The differentiation of fisheries and the enormous impacts on some fisheries on the environment are strong arguments in order to introduce these procedures as valuable supplements with respect to the existing fisheries evaluation and tools for management and also, they are able to provide the appropriate environmental information for all Marine Spatial Planning (MSP) process.

Areas for the nature conservation that are protected with the help of marine ecosystem and to protect the fisheries have different aims. In this way, the legal procedure when developing the marine protected area should rely on whether they are developed for nature conservation purpose or as a fisheries resource management tool.

Fishing in a marine protected area with respect to the conservation should be controlled with the help of conservation law. And, it has been also argued that the marine protected areas (MPAs) with respect to the conservation purposes, in the highest protection category, should primarily be developed as fully protected marine national parks as well as the marine reserves.

Figure 6.1. *An illustration of Tanzania marine conservation.*

Source: Image by Flickr.

All across the globe, conservation conflicts with the fishery interests are very common. A latest example id the failure in order to develop the large Antarctic Ocean Sanctuary. The commission for the conservation of the Antarctic Marine Living Resources, in November 2018, could not agree on the establishment of the sanctuary.

The delegations from Norway, China, and Russia have voted against the proposal to develop three new marine protected areas (MPAs) in order to protect the Antarctic marine living resources, in the Weddel Sea and in the Western Antarctic Peninsula, respectively (ASOC 2018). This is a clear instance of the typical global conflict among marine nature conservation and the industries of fisheries.

As compared with the terrestrial protect areas, some of the MPAs have been developed. Most of these are small in size, and they are allocated in the coastal regions, and also, they are not properly enforced and generally in conflict with the fishery (Barcott, 2011; WRI, 2011; Grip and Blomqvist, 2018).

Conservation and the fishery interests are generally planned and regulated separately, in strict sectoral procedures with the help of the separate organizations, ministries, and legal framework.

Along with that, the two interests are related to different global as well as regional environmental and fishery organizations, which can play a very significant and powerful players in politics (Redpath et al., 2013).

The severe sectoral division of the activities makes the cooperation and coordination more complex, in the context of multisectoral MSP and the process of the decision making, when the MPAs are developed for the purpose of conservation (Rossiter and Lopez-Carr, 2016).

6.2. MARINE NATURE CONSERVATION

Marine nature reserves are every so often developed, having the determination to safeguard, restore, and develop meaningful or significant marine habitats or habitats for the species that are endangered. Designation as well as established fisheries and the livelihoods of the fishing communities (Barcott, 2011; WRI, 2011).

In the matter of fact, fishing, of course impacts the food webs and the ecosystem that an individual want to safeguard (Dayton et al., 1995).

There are always several numbers of controversies occurred among the conservation and the fisheries are most common in the coastal waters at the national level, but also,

They took place regionally as well as all across the globe, having the concern what should be safeguarded in an marine protected area, along with the ecological values or the socio-economic value of the commercial fisheries (Pita et al., 2011; Kearney et al., 2012; Laffoley et al., 2019).

According to the basis of the conflicts between the two interests is that their needs with respect to the same space are generally a challenging task to settle them down. The supposed level of violation with respect to the interest must be addressed with in negotiations among the parties that are involved.

A nature conservation interest objectifying at safeguarding the several habitats and the species, on the other hand, the commercial fishery wants to use or utilize the fish as a resource. The strength of the conflict relies on the strength of the violation and the legislation that is involved.

The fishery is every soften backed up with the help of stronger sector legislation and stronger user organizations. The conservation interest can also help or assisted with the help of environmental organizations and the strong conservation legislation, but there are generally weaker as compared on the fishery side.

Of course, the nature conservation and the fishing interests can interact, but it is very important to acknowledge that the main goal of the MPAs for the means of conservation and fishery resource management are different.

According to the several interviews that are conducted with conservation officers with respect to the management of the marine nature reserves at all coastal CABs in the several regions of Sweden, the necessity for the improved coordination among the nature conservation and the fishing interests was consisted with the help of respondents.

Several numbers of people debated that the fishing in a marine nature reserve should be controlled depending on the conservation law or, if under fishery law, in line with the conservation law (Grip and Blomqvist, 2018).

And also, it should be noted or observed that the experiences of how to regulate fishing in the marine protected region in Sweden as well as other nations, is anchored in and followed with the help of a coordinated work of the regional marine environmental commissions.

6.2.1. The regional marine environmental commissions and fishery

The regional environmental commission (e.g., HELCOM, OSPAR, and UNEP Regional Seas Programme) can cope up with the environmental impacts of fishing, but not the fisheries resource management, which is the responsibility of the fisheries agencies as well as authorities.

In spite of HELCOM, for instance, these issues are discussed within a separate forum the Baltic Sea Fisheries Forum (BALTFISH). And also, in order to provide assistance to the governments with respect to the regulation of the fisheries in Baltic marine protected regions, HELCOM has created a special project: Managing Fisheries in Baltic MPAs (BALTFIMPA).

6.3. MARINE PROTECTED AREAS FOR NATURE CONSERVATION

Marine protected regions have been used for long interval of time, as a conservation tool with respect to the benefit of a specific kind of fisheries.

In the present interval of time, marine protected regions are subject to gain policy attention and are more generally seen as a nature conservation tool, that are primarily focused on safeguarding as well as preserving the several numbers of various marine ecosystems.

The biodiversity as well, with the ecological services they are providing (Hoagland et al., 2001; Laffoley et al., 2019; Maestroa et al., 2019).

Figure 6.2. An illustration of Funafuti marine conservation area.

Source: Image by Wikipedia.

National marine protected regions legislation fluctuates between the nations and are depending to the form of the government, public administrative structures, etc. (Pomeroy et al., 2005; Barcott, 2011). It has been discussed that the development of the marine protected regions, in the context of conservation should having the concentration on the strongest protection category (e.g., marine reserves and marine national parks as well) and be controlled accordingly with the help of the conservation law (Costello and Ballantine, 2015; Costello, 2018).

Fishing in a marine protected area that is developed for the means of nature conservation may be uncontrolled, but is generally controlled in some other way (FAO, 2011). Generally, this regulation is done under the fishery law and jurisdiction (Mesnildrey et al., 2013; Grip and Blomqvist, 2018).

These kinds of regulation are generally based on the negotiation among the two interested parties and relies on the institutional conditions and prevailing politics and on law enforcement organizations that provides assistance in order to implement the process of regulation.

With respect to the negotiations, the institutions are generally having the focus on accomplishing a prosperous fishing with minimal effect on the environment.

The results are generally a situation of compromise, where the fishery is controlled in order to mitigate its impacts on the habitants and the species. Never the less, MPAs are the regions that permits fishing, even though controlled, and cannot safeguard all the characteristics of the biodiversity.

Fishing that claims to be a sustainable process, and is allowed in more than 94% of the all MPAs (Costello and Ballantine, 2015). MPAs are allocated by the means of the conservation or as fisheries resources management tools have various types of objectives and in this way, their management should be kept legally separated under the conservation and fishery jurisdiction, respectively (Hilborn, 2016).

Management of the MPAs, with respect to the different laws, under the different governments and supervisions, each eagerly defending its mandate, and they are easily leading to target conflicts.

Strictly protected conservation of MPAs (no fishing allowed) according to the conservation law, as, for instance, marine nature reserves in New Zealand, are not common (Enderby and Enderby, 2006).

The positive long-time experiences of strictly protected marine nature reserve in New Zealand, that are based on the socio-economic as well as ecological factors, should be applicable all across the globe and deserves to be applied or at least tried extensively (Ballantine, 2014).

And also, strictly protected areas have every so often proven to be very advantageous with respect to the fishing outside the MPAs (Costello, 2018).

6.4. PROBLEMS AND ISSUES IN FISHERY MANAGEMENT

In the present interval of time, the interests in MPAs depicts the dissatisfaction with the typical methodologies or approaches with respect to the conservation of the marine ecosystems, more particularly fishery management, which every so often have failed in order to fulfill the objectives of the society for the sustainable use of the marine resources and the protection of biological diversity and the productivity as well.

Overfishing shows one of the most difficult tasks or difficult issues in the context of marine conservation. With respect to the global basis or global stats, near about 44% of the fish stock of all across the world are now entirely exploited, and near about 25% of them are overexploited and clearly in need of the urgent conservation and the management measures (Gracia and Newton, 1997; FAO, 1999).

Figure 6.3. Problems and issues in fishery management.

Source: Image by Needpix.com.

Collapse or the enormous decrement of the marine fish stocks, for instance, the Atlantic cod (*Gadus Morhua*) off Newfoundland and on Georges Bank, call attention to the potential for the failure outcomes from the restrains of the fishery science and of the current management system (Walters and Maguire, 1996; Fogarty and Murawski, 1998).

The view of fishery scientists is that it was at some level a failure of science that gave birth to the collapse of the Newfoundland cod. The scientific advice that was given to the managers or the executives was not always correct or timely, and the management failed to act in time on either the mistaken advice or the corrected advice.

In the several regions of the United States, the overexploitation of the bluefin tuna and of mixed-species ground fish stocks convincingly shows that the results of the overcapitalization and the excess effort.

Issues that are linked with the traditional fishery management that harming the sustainability of the marine fisheries have been highlighted in the latest reviews (e.g., NRC, 1999a). overfishing and the attendant fishing mortality rates are much higher and badly controlled lead the list. Overcapacity, the existence of the several numbers of participants or units of efforts in fisheries, is a respective as well as critical issue.

Open access to the fisheries tends to favor overexploitation instead a stewardship, a typical impairment to the efficient management (NRC, 1999b), never the less, very few of the commercially important fisheries in the several regions of the North America and Europe are entirely open access in the context that most of them are controlled, and some of the have restricted allowance, and several numbers of them have restrictions on effort.

In addition to this, another issue is the failure of the management to act expeditiously and conservatively or to respond suitably. Approaches to the management that are based on the ecosystem (e.g., NMFS, 1999) have improved from a concern about by catch, habitat destruction, and the failure in order to well thought out the significant biological interactions (e.g., predator, and its prey).

Shortfalls is an ability of the scientists that is used to harvest precise stock evaluation have at time provided poor advice to the executives or managers (NRC, 1998b). Stock assessment or stock evaluation and the resultant management measures always are consisting of a level of uncertainty.

In order to be very productive or efficient with respect to this uncertainty, necessitates that the assessment or evaluation be interpreted conservatively, in such a way that stock size is not overestimated and subsequently overfished. Conservative, adaptive, as well as flexible approaches can compensate for the uncertainty of stock assessments, but frequently these kinds of aspects or features are lacking from the traditional management.

Overfishing is in large part an outcome of the extra efforts and the capacity in fisheries (NRC, 1999a). every so often, the executives of the fisheries have been not able to regulate or manger the fishing efforts, as an outcome in the unsustainable level of catch. This has been a particular issue with respect to the open access fisheries where the management does not restrict the number of the participants or high individual efforts.

With respect to this setting, the economic incentives favor short-term exploitation over long-term sustainable use because the economic advantages or benefits of sacrificing the current catch in order to build the stock again are comparatively intangible as compared to the short-term requirements (bills to be paid), and long-term benefits may have to be shared with the newcomers when the fishery recovers (Hillborn et al., in press).

As more people enter the industry of fisheries or to enhance their capabilities of fishing, the future yield to the individual fisher reduces. This, every so often, fosters the competition, in order to maintain or even elevate the individual catch levels even as stock reduces.

In response, managers or executives may shorten the season of the fishing, participants then elevate their fishing power, and effort becomes concentrated in time, and every so often, as an outcome, in "races for fish" or "fishing derbies."

6.4.1. Conventional fishery management

Generally, the traditional fishery management was high maintaining cost, yet sustainable, produces with the help of the numbers or weight of fish caught, the size of the fish caught, or the time and space or region within the fishing is permitted. The intent in each case is to regulate the fishing mortality rates.

The traditional methodologies with respect to the fishery management in the United States can be briefly categorized with the help of three main elements:

1. An underlying the fisheries science and the management patterns;
2. A group of traditional management tools; and
3. The fishery management system.

Figure 6.4. Traditional fishery management.

Source: Image by Wikimedia.

6.4.2. Fishery paradigm

Fishery management depends on the approximation of the size of the population of specific types of species in order to find out how many fish or what fraction of the biomass of the population can be caught without damaging its reproductive potential.

In order to make these kinds of determinations, the management relies on a conceptual model of a fishery that makes three simplifying assumptions:

1. The fishing fleet targets and the exploits a single-species stock;
2. The stock of the interest is separated temporally or spatially from the other stocks; and
3. The individuals are accurately mixed, in this way, the effects of fishing are well spread over the whole stock.

These kinds of expectations, which are far from true in most of the situations, can have the critical results for the efficiency of the fishery management.

6.4.2.1. Single-species, one stock

Most of the management measures are conducted at individual stocks of a single species and do not take into the account species interactions, for example, the relationship of predator and prey. A basic assumption of the most models.

It has been observed that management measures are directed at individual stocks of a single species and these measures do not take into account species interactions, like predator-prey relationships. A basic assumption of most of the models that are used to determine a catch level is that the catch rate a stock can sustain can be designated that is based upon the average productivity of the stock.

In turn, the term productivity is presumed to rely mainly on the size of the adult stock. In this scenario, the control of adult stock size is the main means of ensuring the sustainability of the fishery. In addition to it, stocks are expected to respond in a density-dependent manner and thus, these stocks are assumed to have maximum productivity at intermediate stock sizes.

In this way, the maintenance of the stock size that permits maximum sustainable yield (MSY) historically has been a primary management goal. In addition to it, fishing at a rate that produces MSY on average over good as well as bad years has been the target.

In fact, it has been observed that fishing at the MSY level (that is a fixed exploitation rate policy) does not ensure a stable adult population size or constant catches in the future. This is because of substantial variability in reproductive success and recruitment.

More than two decades ago, it was recognized (Larkin's, 1977, famous epitaph on the concept of MSY) that it is thus too risky to set a constant quota for catch at MSY.

It has also been observed that in good years, fishers may flourish with MSY-based catches, but in the time period when the environment is less favorable and recruitment and productivity decreases, the stock will reduce and also, MSY may rapidly lead to overfishing.

The fishing rate that corresponds to MSY still continues to be a criterion in determining the optimal yield, the regulatory target that is used to manage marine fish stocks in the United States (NOAA, 1996a). Although, in the present times, FMSY is viewed by many as a threshold that should not be exceeded, rather than it is viewed as a target at which to aim.

Also, it has been observed that in the present times, more conservative quotas and exploitation rates are recommended, due in part to recognition of the limited ability to estimate as well as implement F MSY or, for that matter, other target fishing mortalities via catch or effort control.

As Hilborn and Walters (1992) noted that obtaining an estimate of MSY (or F MSY) generally needs fishing at levels that already exceed it.

However, it is very clear that management must be tailored to the life histories of the individual species. The individual stock, as well as the single-species approaches to management, are not effective for multispecies or even for mixed-stock associations. In this, many different species or stocks which are having similar prey requirements and habitat overlap in their ranges.

Some of the good examples are reef fish off the south-eastern United States and other example is rockfish in the northern Pacific Ocean. It has been observed that warm-temperate species like gag (*Mycteroperca microlepis*), red hind grouper (*Epinephelus guttatus*) and scamp (*Mycteroperca phenax*), for example, co-occur to such an extent that catch restrictions that are placed on one species in the complex generally result in increased regulatory discards while fishing for associated species.

This is true for all the 83 species of rockfish that are managed as a complex off the Pacific west coast. Certainly, regulatory discards will increase the rate of mortality of the restricted species and it also threatens its recovery.

6.4.2.2. Individuals Are Perfectly Mixed

The primary element of the fishery management paradigm is defined as the concept of a well-mixed stock. Migration patterns, as well as more general spatial processes, are considered as the fundamental components of fishery science. The proper information of spatial processes serves to delineate or outline management units. Each is viewed as a "dynamic pool" which is isolated from the rest.

It has been observed that most theory and management have been conceived for large-scale, commercial fisheries that target comparatively mobile species (e.g., plaice, tuna, gadoids). For this, the dynamic pool assumptions may offer a reasonable simplification, at least at the scale of a fishing ground.

In this conceptual model, the use of spatially explicit approaches to manage every single unit has been largely missing (Shea et al., 1998). This is because the effects of fishing are "diluted" in the pool. In this conceptual model, there are dominated marine fishery management.

As a result of this, it has been used extensively to low-mobility species for which the paradigm is clearly not appropriate. Some of the examples are rockfish (*Sebastes* spp.) and sedentary invertebrates. Fishing effort on comparatively sedentary species specially targets the highest fish concentration. This results in a mosaic of fishing mortalities and it also results in serial depletion of fishing grounds.

It has been observed that management by conventional means is complicated. This is because the dynamics of the fishing process, resource, and monitoring are dominated by local processes. Often, that is not possible or it can be said that it is impractical to assess. For instance, in reef fisheries, the relevant scale for assessment as well as management may correspond to a single reef.

The other tool that is used to limit fishing effort is control of access to the fishing grounds. This strategy is called "spatial management." This strategy can be used as either temporary (that is seasonal or year-to-year) or it can also be applied as permanent closure of portions of the fishing grounds.

Spatial management can be helpful in order to control access to resources. Also, it has been practiced for centuries (Cushing, 1988), but it has not played an important role in the management of marine fisheries. In the present times, its importance in the management of benthic shellfish is accepted (Orensanz and Jamieson, 1998; Perry et al., 1999). Also, the novel management schemes that include area rotation are also accepted (Bradbury 1990, 1991; Perry et al., 1999).

However, it has been observed that there are many species that could benefit from spatially explicit management. These species are mainly those which are having a relatively stationary adult stage. A main constraint on the execution of spatial management is the absence of spatial catch data for many species and also, the complexity of spatially explicit stock assessment models.

Even though spatially explicit components generally are missing, some of the existing fishery models do recognize that variables other than just the adult stock size also have impact on the productivity. However, these variables are generally are treated as random variations beyond human control.

Therefore, management emphasizes on regulating the size of the catch and the attempts are directed at acquiring it, while it has been observed that the environmental factors are downplayed. Persistent trends or the differences in productivity that is driven by environmental conditions should be considered in policy evaluation even when not controllable.

In addition to it, environmental variables that have impact on the quality of habitat and are directly affected by the activities of human are now addressed explicitly in fishery management plans (FMPs). This is following the reauthorization of the Magnuson-Stevens Fishery Conservation and Management Act (MSFCMA, NOAA, 1996a) and its primary focus on the important fish habitat (EFH).

This emphasize on habitat has started a shift toward spatial considerations and designation of areas that is essential for the productivity of the species that are economically important.

These designations, of important fish habitat and habitat areas of particular concern, have helped in boosting the interest in protected areas and it also supported the use of MPAs. This is used as a legitimate tool for fishery management.

It has been observed that long-term, large-scale closures have been introduced as single-species refuges from fishing to encourage rebuilding of depleted stocks. In the context of several numbers' respect, they resemble permanent reserves, except that such areas may revert to their former status when restoration is attained.

It has been observed that long-term closures are becoming more common as a population-rebuilding tool (e.g., northern cod stock off Newfoundland). In this, depleted populations may need many years in order to restore. Long-term closures may attain some goals of reserves, even though the advantages may be transient if subsequent fishing mortality cannot be controlled.

The other form of spatial management is the application of rotating fishing areas. In this, just a fraction of the fishing grounds is opened in a given season, the rest of the fishing grounds is being closed for specified periods (often years) in order to promote growth of young animals. It also permits them to reach more valuable sizes. The area that is opened is rotated from year to year.

This methodology integrates the temporal as well as spatial closure in order to control the fishing. In addition to rebuilding the fish stock, rotating closures can permit the habitat and the biological communities having the means to recover from the effects of fishing, for example, damage to bottom habitats by trawling and movement of "fixed" gear. Never the less, recovery of the habitat and biological communities can necessitate closures on the order 5 to 10 years (Collie, 1997).

6.4.3. Conventional management tools

Management that is based on the fishery patterns above centres on measures that regulates the activities of the fishing and the level of catch, as a substitute to measure that directly motivate the management of the habitat or consideration of the environmental variables that are influencing the productivity of fish.

Management that is based on the fishery patterns above centers on measures that control the activities of fishing and the level of catch, as an alternative to measure that directly promote the management of the habitat or consideration of environmental variables that are impacting the productivity of fish.

Generally speaking, the main objective is to manage the exploited population, for example, that they are maintained at productive levels (close to or above MSY) that provide assistance to a high but sustained fishing yield and in order to do that, it requires rebuilding plans when yields fall below a minimum stock size threshold.

Figure 6.5. An illustration of traditional fishing boats: Conventional management tool.

Source: Image by Wikipedia.

Typical management methodologies, in order to control the exploitation rates fundamentally depends on placing the limits on the amount or efficiency of fishing effort (which is known as effort-controlled fisheries) or on the total amount that are caught with the help of specifying the catch quotas and the allocations as well (which is known as quota-based fisheries).

In both of the cases, a target fishing rate is first specified, whether it is a constant or it is a variable, with respect to the stock condition, according to the analysis of the historical experience with the fishery, on the experience with the fisheries for same kind of species, or on modeled responses of the fishery in order to stimulate the fishing mortality.

6.4.4. Effort controls

There are several numbers of forms of effort controls, which is consisting of vessels, numbers of fishers, gears, and time fished. Generally, these are usually the first controls applied to a fishery in order to slow down the rate of catch. Gear restrictions can be consisting of the amount, type, or dimensions of the gear or particular features of the gear, for example, size of the net mesh, configuration of the fish traps, and hook spacing on long lines.

Vessel restrictions can be consisting of the horsepower of the engine, length of the engine, and the design of the engine. The several numbers of fishers can be controlled with the help of allocating the licenses to either fishers or vessels.

Time fished can be controlled with the help of constraining the amount of the time that is available for fishing with the help of seasonal closures, "day-at-sea," or particular days and hours when the fishing is permitted.

In order to execute a target fishing rate with the help of effort regulations, executives or the managers require a reliable approximation of the fishing mortality that is caused by each unit of fishing effort to be permitted, a parameter known as catch ability.

The approximation of the catch ability is then utilized in order to find out the amount of the fishing effort (e.g., the number of the total days spend at the sea to be allowed) which is chosen exploitation rate or fishing mortality, in such a way that:

F_{target} = catch ability × effort

Generally, the parameter 'catch ability' is assumed to remain constant as size of the stock fluctuates. Never the less, catch ability can differ. For example, fishers can elevate their effectiveness or productivity when the size of the stock decreases, and this could, as an outcome causes the overexploitation, a typical reason for the failure of the traditional management.In addition to this, in order to measure the attempt to control the rate of exploitation directly, other forms of the effort control can be executed in order to decrease the fishing power, protect the vulnerable life-history stages, or increment in the market value of the fish.Temporal closures, for example, are regularly utilizes in order to safeguard the fishery resources at times when they are specifically vulnerable to fishing, for example, when fish aggregate on spawning grounds.And also, temporal closures can be utilized in order to allocate the fishing over the season, in such a way that it increases the value of landing. For example, fishing might be prohibited through the course of the parts of the year when the population is composed of small or poorly conditioned fish. Closures of Dungeness crab (cancer magister) fishery through the course of the molting season, when the quality of the meat is poor provides one instance (Methot, 1986).

Extensive usage of the temporal closures to safeguards the entire communities or several numbers of species and the habitats is less common, but perhaps of greater relevance in order to develop or create marine reserves as an ecosystem approach with respect to the fishery management.

6.5. FISHERY MANAGEMENT SYSTEMS

In the several regions of the United States, the management of the fisheries generally is undertaken at geographic scales that ranges from local to national. Allocation of the responsibilities and the execution of the efficient management is a very complicated task to do. Jurisdiction of responsible institutions and the organizations can overlap in some of the fisheries industries.The eight regional fishery management councils (NOAA, 1996a) have primary responsibility for the management in the U.S. Exclusive Economic Zone (EEZ), but they may share responsibility with other regional management institutions for the coastal migratory species, more particularly those that took place in the nearby shore and estuarine regions of the coast.

For instance, the Mid-Atlantic Regional Fishery Management Council share the responsibility with respect to the management of the coastal species, for example, bluefish (*Pomatomus saltatrix*), weakfish (*Cynoscion regalis*), and the summer flounder (*Paralichthys dentatus*) with the Atlantic States Marine Fisheries Commission, which shows the interest of the state in migrating the species that are fished in the coastal zones and the estuaries of the Atlantic Coast states.Management systems are even more complicated for these kinds of species, cause state agencies also are busy in the regulatory process within their jurisdictions.

Figure 6.6. Fishery management system.

Source: Image by Obama White House.

In addition to this, the management of these kinds of species is directed at additional regional levels (e.g., the Chesapeake Bay, in which the states of Virginia, Pennsylvania, and Maryland; the Potomac River Fisheries Commission; and the District of Columbia exercise jurisdictional control).

Generally, the management system is consisting of several numbers of output as well as input controls in order to regulate the fisheries over their geographic ranges. Quota allocations, every so often among the sectors of the fishery (e.g., commercial, and the recreational), are common; minimum sizes or the other size regulations may apply.

Constraints on the gears, seasonally closed regions, seasons, and the integration of methods, every so often with the specific geographic regulation within the range of the targeted species, are the tools that the managers generally apply. Not only is it difficult to acquire the agreement in order to control the resources, but the success of the management measures is every so often uncertain.

The uncertainties with respect to the success of the management systems lie in the attendant uncertainties that generally categorizes the science and the management of the fishery resources. The science of stock assessment itself can be uncertain for several numbers of fished stock.

Political pressures on the executives or the managers and the institutions can dictate the management policies and the responses, every so often to the disadvantages of the long-term benefits to fisheries.Disputes among the sectors of the fisheries, for example, different gear users, or recreational versus commercial fishers can dominate the dialogue and sometimes as an outcome depicts the compromises that do not make the best policy for management.Faced with uncertainty in science and social conflict, managers historically have been slow to act in order to conserve the fishery resources. Legislation, such as the national standards of the MSFCMA (NOAA, 1996a), present, at least to some, conflicting objectives of the conservation, social interests, and economics that slowdowns or misdirect management actions. In the end, in order to be completely efficient, the management systems must encourage compliance, either with the help of enforcement or with the help of providing proper institutional incentives to comply with regulation. As discussed in the above topics, the management systems that confer user rights and the participation of the shareholders in the management process can enhance the compliance.As it has been observed, management in most of the U.S. EEZ is controlled with the help of regional management councils (NOAA, 1996a). In the present interval of time, there are thirty-seven FMPs submitted with the help of regional councils and approved by the Secretary of Commerce.

The eight councils have jurisdiction overbroad, discrete geographic regions, never the less, sometimes their management authority is shared for migratory species. Jurisdictional problems may be meaningful for migratory stocks, more particularly coastal stocks that cross the boundary between state and federal waters at three miles from shore (for most states).

6.6. ISSUES IN EVALUATING REGULATORY SYSTEMS

Over the time period of 45 years, Michael Graham (1949, The Fish Gate) discoursed that in the great law of fishing.

Fisheries that are unlimited become unprofitable. When the value of the catch is high and the cost of harvesting is low, biological overusing is likely to occur, that is when the net value expressed as resource rent is great.

The bionomic equilibrium, which is defined as the state at which the value of the sustained catch is no greater as compared to the cost of harvesting can be impacted with the help of several numbers of factors for example increase of catch or lowering of cost or changes in quality of technology inputs and so on.

In this way, safeguarding of the fish stock and the production is only one part of the problem. Being common property resource there is always a tendency to over develop fishing capacity to extra capacity arose to the several numbers of issues.

Regulatory systems to direct, these will need periodic review and evaluation. Pear (1980) has outlined the following which appear appropriate or suitable with respect to any regulatory system that can adapt:

Thus, the protection of the fish stocks and the yield is only part of the issue. Being common property resource there is always a tendency to over develop fishing capacity to excess capacity giving rise to many problems.

Regulatory systems to govern these will require periodic review and evaluation. Pears (1980) has outlined the following which appear relevant to any regulatory system we may adopt:

1. Its efficiency in regulating the fishing capacity;
2. Its effect on the fishing technology;
3. Its adaptability and the robustness in the face of inevitable modifications with respect to the fishing technology, price of fish, cost, and the availability of the resources;
4. Its effect on the distribution of effort;

5. The way it will allocate the economic gains from the rationalization between the fishermen, the Governments, vessel owners, as well as others;
6. The level to which it would produce dislocation and impact the employment; and
7. The administrative complications and the cost.

6.7. REGULATING FISHING REPORT

Controlling the fishing efforts revolves around safeguarding as well as improving the productivity of the stock with the help of controlling the size of the composition of the catch.

The value of the sustainable production or the size of the fish in the catch relies on the selectivity of the gear, location, and timing of fishing, introduction of minimum size of mesh, closed seasons and the restricted fishing areas.

6.7.1. Closed season

Closed season has several numbers of various connotations. First is the banning of the fishing through the course of part of the year to specifically safeguard the life history stages of fish stocks at the time of spawning, juvenile stages, and larvae. Second reason is, closed season is applied when the CDPUE declines to a certain point.How this could be applied to a multi-species fishery is complicated and will need location particular observations and decisions.

Closed season according to Pears (1980) 'simply reduce time in which fish may be taken, without affecting a long-run reduction of the effort. For the individual enterprise, fishing through the course of the shortened season will become more urgent and this will motivate the distortions in the design units with the help of providing incentives for the increased speed and the storage capacity.Any short-run increase in the profitability will additional entrants. As the fishery adjusts to the shortened season, the season will have to the progressively shortened in order to avoid the total pressure from the increasing.If the fish are equally available through the course of the closed as well as open seasons this process will as an outcome in no reduction in excess fishing capacity.On the contrary, the fleet will expand to the size that are capable of taking the catch in a shorter season. In short, the seasonal closures, such as gear constraints, are ineffective means of the regulating the fishing effort.

6.7.2. Conservation

With respect to the biology, the conservation issue is ubiquitous. On the other hand, safeguarding the stocks, it objectifies at also providing the sustainable yield.In multi-species fisheries it is difficult to anticipate the decrease in the recruitment because of the fishery dependent factors. One way out is to control the total fishing effort applied to the society. This could avoid the modifications in the species composition which are economically detrimental or irreversible.

In several regions of Malaysia, Indonesia, and Thailand, the tendency has been for a shift from the predator species to the prey species. This has had a twofold threat to the predator species, with the help of the threat to the predator species from below normal recruitment and secondly the decrement in the forage food supply of low value fish.This depicts to the point out complications with respect to the conservation and the number of issues that are inter-related in developing the conservation criteria for the fish stocks and the communities. In these kinds of cases, Pauly, and Murphy (1982) opine that "the biological criteria for the conservation can be summarized in the simple idea that the modifications in the community or the target species abundance should be reversible on some of the reasonable time scale."The regulations that are based on the conservation may have to cope-up with the social values, economic performance, more particularly equity, administrative feasibility and the political acceptability.Unhealthy competition and the conflicts among the industrial fisheries operating off-shore with the trawlers and the artisanal fisheries constrained to a narrow coastal strip, which results according to the Sardjono (1980) in the dramatic fishery regulations and the banning of the trawling in the Indonesian Waters. This is an extreme case and shows the level to which conservation problems are taken up in order to protect the resources for sustainable yields and also fulfil the requirements of social comments.

6.7.3. Economic performance

The bio economic equilibrium is escorted with the help of increment in the economic inefficiency when the fisheries are uncontrolled. Revenues from unit tends to decrease when more fishing effort is attracted to the fishery.

Bio economic equilibrium is reached in theory when the lack of profits discourages the further entry. The low income prevalent in the fishery can also attract controlling.Bain (1984) remarks that the crateful attention to both cost as well as benefits because of the regulation is needed, if the fisheries are to be a source of value and not a burden to a country.

One area Bain questions is whether the modifications in the fishery regulations could take into the direction towards the increment in the foreign exchange or whether they could be a drain on these kinds of funds.

6.8. CONCLUSION

All across the globe, there has been a surge of interest in allocating the regions of the seas as marine reserves and the protected areas in order to maintain and conserve the marine species and the habitants that are threatened by the activities of human. There is an increment in the consensus that living marine resources necessitates more stringent protections.

Crises facing several numbers of marine ecosystems that are increasing and attracting more and more attention of the public. Among these are the latest collapse of the Newfoundland cod fishery, the near-collapse of the ground fish fishery in New England, and the loss of coral reef communities to disease and overfishing.In this way, these is an extensive concern between the scientists, policy-makers, and the public at large about the current status and uncertain future of marine ecosystems. Better approaches with respect to the utilizing and safeguarding the living marine resources are required.

Never the less, selection of the best methods in order to maintain or restore the health of the marine ecosystem is very difficult job for resource managers and a source of disagreement between the user groups, scientists, and the conservation community.An individual does find general agreement, never the less, on the shortcomings of most of the current management policies. For instance, traditional fishery management generally having the focus on single species and is concerned with the particular types of problems.For example, MSYs, suitable fishing mortality rates, effects of the seasonal closures of certain fishing grounds, and suitable size limits in order to safeguard the juveniles and the spawning potential of the stock.

Whether or not these strategies with respect to the single species management accomplish their particular objectives, their practice every so often neglects other important and pervasive issues.

In addition to this, controlled designed for on fishery may impact other fishery in a bad way, on the same fishing grounds with the help of gear conflicts, habitat destruction, bycatch, or subtle but important shifts in the relationships of the predator and prey.

REFERENCES

1. Arti, M. J., Abhay, D., Ejaz, P., & Kadri, R M., (2018). *Sustainable Management of Marine Fisheries ReSource: An Overview*. [online] www.researchgate.net. Available at: https://www.researchgate.net/publication/324861546_Sustainable_Management_of_Marine_Fisheries_Resource_An_Overview (accessed on 8 March 2020).
2. Beddington, J., Agnew, D., & Clark, C., (2007). Current problems in the management of marine fisheries. *Science*, [online]*316*(5832), 1713–1716. Available at: https://science.sciencemag.org/content/316/5832/1713 (accessed on 8 March 2020).
3. Beddington, J., Agnew, D., & Clark, C., (2007). Current problems in the management of marine fisheries. *Science*, [online]*316*(5832), 1713–1716. Available at: https://www.researchgate.net/publication/6248405_Current_Problems_in_the_Management_of_Marine_Fisheries (accessed on 8 March 2020).
4. Grip, K., & Blomqvist, S., (2019). Marine nature conservation and conflicts with fisheries. *Ambio*. [online] Available at: https://link.springer.com/article/10.1007/s13280–019–01279–7#Abs1 (accessed on 8 March 2020).
5. Islam, M., Shamsuzzaman, M., Hoque Mozumder, M., Xiangmin, X., Ming, Y., &Abu, S. J. M.,(2017). Exploitation and conservation of coastal and marine fisheries in Bangladesh: Do the fishery laws matter? *Marine Policy*, [online]*76*, 143–151. Available at: https://www.sciencedirect.com/science/article/pii/S0308597X1630553X (accessed on 8 March 2020).
6. Mora, C., Myers, R., Coll, M., Libralato, S., Pitcher, T., Sumaila, R., Zeller, D., Watson, R., Gaston, K. & Worm, B., (2009). Management effectiveness of the world's marine fisheries. *PLoS Biology*, [online]*7*(6), p. e1000131. Available at: https://journals.plos.org/plosbiology/article?id=10.1371/journal.pbio.1000131 (accessed on 8 March 2020).
7. Serge, M. G., & Ignacio, D. L. M., (2001). *Global Overview of Marine Fisheries*. [eBook] Reykjavik, Iceland: Food and Agriculture Organization. Available at: http://www.fao.org/fishery/docs/document/reykjavik/pdf/01garcia.pdf (accessed on 8 March 2020).
8. Silas, E.G., (n.d.). *Conservation and Regulation in Marine Fisheries*. [eBook] Kerala, India: E.G. Silas Foundation for Nature Conservation.

Available at: http://eprints.cmfri.org.in/7830/1/685._Sil.pdf (accessed on 8 March 2020).

9. Tait, M., Dale, S., Ray, H., & Quentin, G. R., (2010). *Marine Conservation and Fisheries Management: At the Crossroads*. [online] www.researchgate.net. Available at: https://www.researchgate.net/publication/288847099_Marine_conservation_and_fisheries_management_At_the_crossroads (accessed on 8 March 2020).
10. The National Academics of Science Engineering Medicine, (2001). Marine Protected Areas: Tools for Sustaining Ocean Ecosystems (2001) Chapter: Conventional Management of Marine Fisheries. [online] The National Academics of Science Engineering Medicine. Available at: https://www.nap.edu/read/9994/chapter/5#39 (accessed on 8 March 2020).

7

Artificial Reef in Fishery Management

CONTENTS

This chapter is about the artificial reefs that are developed by the human being to have a good amount of catch inside the sea near seabed. This chapter explains the importance of artificial reefs, their procedure of making them in the sea near sea bed and all the criteria that are required to be kept in mind while putting artificial reefs. There are different types of artificial reefs that are present. These artificial reefs are selected according to the type of fish to be catch and according to the area.

7.1. INTRODUCTION

Artificial reefs are used to attract the fishes and facilitate its capturing for human consumption. This technique is the commonly used technique around the world. There is an evidence in the Mediterranean Sea that shows the first artificial reefs was accidentally created in around 1500s. At that time, the rocks were used to anchor. They use tuna fishing nets which were left on the seabed at the end of each fishing season.These anchors were then accumulated over the time and then created new rocky habitats that became inhibited by the fish. These rocky habitats were exploited by the local fishermen at the time of tuna fishing season (Riggio et al., 2000). Through the history, the similar practices are being employed by the artisanal fishers (Simard, 1995) across the world. The modern concept of the artificial reef was first given by Japan during 20th century after the World War second. This has been adopted in the second half of the 1900s in the Mediterranean. The increasing interest for the artificial reefs gave the concern that was regarding the possible negative impact that is caused using unsuitable material and dumping waste.

Figure 7.1. *Artificial reefs for fish.*

Source: Wikimedia Commons.

7.2. DEFINITION OF ARTIFICIAL REEFS

It is important to have a common understanding about the word artificial reef. An Artificial reef is a man made under water construction that is typically built for the purpose of promoting marine life in the area of generally featureless bottom.

The definition can be also given as, an artificial reef is a submerged structure which is deliberately placed on the sea bed to impressionist some function of a natural reef. These functions can be protecting, regenerating, concentrating or enhancing population of the living marine resource.

Figure 7.2.*An overview looks of artificial reef.*

Source: Image by Wikimedia Commons.

The placement of artificial reef helps in protection and regeneration of the habitats. The habitat that functions as a part of the natural ecosystem which in return brings no harm. This term basically excludes artificial islands, cables, platforms, mooring, pipelines, and structure that are for the coastal defines like breakwaters, dikes.

These structures are being primarily constructed for other purposes. They are also used for fish aggregation devices (FADs) employment that merely attract fish in certain fishing areas.

7.3. INTERNATIONAL LEGISLATION ON ARTIFICIAL REEF DEPLOYMENT

Artificial reef deployment is an activity that is being covered by many of the international legal instruments. These international legal instruments are those who are protecting the sea against the pollution which occurs due to dumping of unsuitable materials like plastics and many other materials. List of these instruments is being provided below in a chronological order.

The convention that is done for preventing the marine pollution which is majorly caused dumping of waste and other materials also known as 'London Convention' is being bought in the year 1975. This convention was one of first worldwide conventions that was protecting the marine environment from human activities that was making water full of pollutants.

In the year 1996, the protocol that is used for convention on the prevention of marine pollution by dumping of waste and other matter commonly known as the 'London Convention Protocol'.

This protocol was adopted to update the London Convention by incorporating the principles of international environmental law which is laid down at the United National Conference on

In this protocol, Art 4 is being stated that contracting parties "shall prohibit the dumping of any waste or other matter with the exception of those listed in Annexure 1." These exceptions are:

- dredged material;
- Sewage sludge;
- Vessels and platforms or other man-made structures at sea;
- Fish waste or material resulting from industrial processing operations;
- Organic material of natural origin;
- Inert, inorganic geological material; and
- Bulky items primarily comprising iron, steel, concrete, and similar unharmful materials. For these items, the concern is physical impact and limited to those circumstances. Circumstances where such wastes are generated at locations that is having no practicable access to disposal options other than dumping.

In the year 2006, the London Convention Protocol is said to enter into the force. Some specific guidelines are being developed in the year 2008 for the placement of artificial reefs. These guidelines are being developed within the context of the London Convention and Protocol/ UNEP, 2009. These guidelines were approved by contracting with the parties to the London convention and London Convention Protocol. These guidelines are made to aim that supports countries in assessing the proposals in order to do the placement of artificial reefs.

This placement of artificial reef is done based on scientifically sound criteria. These proposals are not legally binding. The United Nation Convention Law of the Sea (UNCLOS) is silent about the artificial reefs.

Even though there is no specific provision made that can address artificial reefs in the UNCLOS. Whereas, this convention specifies the rights that are enjoyed by the states over marine waters. This means that according to the given area of the sea and depending upon the jurisdiction of that area, state might install artificial reefs. This happens in a specific area of territorial waters. These areas are where the coastal state enjoys the sovereign rights.

In the year 1992, the convention on the protection of the black sea that was against pollution which was known as the Bucharest Convention, that was being ratified by the six black sea riparian states. This protection act entered into the force in the year 1994.

The main aim behind these conventional protocols is to protect the marine environment of the black sea. Which simply relates with the preservation or protection for the living marine resources.

The Bucharest Convention have following four protocols:

1. The first protocol on the protection of the Marine Black Sea Environment against the pollution from the Land base sources commonly known as LBS Protocol. This protocol was adopted in the year 1992 and came into the force in the year 1994.

There was a revised protocol on the Protection of the Black Sea Marine Environment against the pollution that was from the Land sources and from the activities. This protocol was adopted in the year 2009 and is still in wait for its force.

2. The second protocol was on cooperation in combating pollution of the Black Sea Marine Environment. This pollution was due to oil and other harmful substances in emergency situations. This is also known as Emergency Response Protocol. This protocol was adopted in the year 1992 and came into force in the year 1994.
3. The third protocol was on protection of the Black Sea Marine Environment against the pollution that was caused by dumping.

This protocol is also known as dumping protocol. This protocol was adopted in the year 1992 and later it came into the force in the year 1994.

4. The fourth protocol was on the Black Sea Biodiversity and Landscape Conservation. This protocol was on the convention on the protection of the Black Sea against the Pollution. This protocol is also known as CBD Protocol. This protocol was adopted in the year 2002 and it is yet awaited to come in the force.

7.4. ARTIFICIAL REEF AREAS

After the first scientifically planner reef that is being installed in the year 1974 off the Adriatic coast of central Italy (Bombace, 1989; Fabi et al., 2011), after this, there were around 250 more artificial reefs area which were created in the 10 Mediterranean countries.

Artificial reef was planned as a fishery management tool. This management tool was expected to increase the fishing yields and to reduce the conflicts that are among different fisheries sectors. The management was also used as a protection device for a coastal sensitive habitat.

Figure 7.3. Image of an artificial reef constructed underwater.

Source: Image by Wikimedia Commons.

Pyramids that are made up of concrete blocks are considered as the most common submerged artificial structure. In order to fulfil the different specific purposes, there are many other manmade structures that are used (Fabi et al., 2011).

There is an involvement of few Mediterranean artificial reefs that have been created for the research purposes and for the recreational fishing and diving.

7.5. OBJECTIVES OF ARTIFICIAL REEFS

Artificial reefs can be considered as the intervention of engineering technology that is done to recover or improve the natural habitats with increasing the productivity and managing of aquatic resources.

In this situation, artificial reefs are used in the coastal waters for many purposes around the world. Some of the purposes are explained below:

- restoring depleted habitats;
- improving populations of aquatic organisms by providing shelter for juvenile and mature individuals as well as for adults during delicate life stages (e.g., moulting season for crustaceans);
- creating potential networks of marine protected areas (mpas) to manage the life cycles of fish and connectivity;
- providing a mean to manage coastal activities and reduce conflicts;
- protecting sensitive habitats from fishing activities;
- enhancing professional and recreational fisheries;
- creating suitable areas for diving;
- providing new substrates for algae and mollusc culture;
- mitigating habitat loss;
- enhancing biodiversity; and
- Research and educational activities.

The objective behind the development of artificial reef deployment is not mutually exclusive. As artificial reefs can be often created for more than one purpose.

The purpose can be protection from fishing and finfish enhancement. For this case they are defined as a multipurpose artificial reef.

7.6. PLANNING OF ARTIFICIAL REEFS

7.6.1. Pre-construction phase

Pre-construction phase is a phase that involves all the action that is needed to be undertaken in reference to the installation of an artificial reef. This installation take place from the decision of constructing the reefs to the submission of the plan to the authorities those are more knowledgeable or skilled.

The first main step that is taken while planning an artificial reef consists a broader goal that focuses on its construction and as well as in evaluating the ecosystem. The ecosystem is where artificial reef is deployed and how the environment will get affected by the process of immersion of new substrates.

This does not happen during the immediate installation process but it also happens over the reefs intended design life.

There are many questions that are related to the effect of the artificial reefs after the installation that can be addressed to the effective use of pre-installation environmental assessments. This basically investigates the anticipated impacts that can be both positive and negative of the installation of the reef and purpose of the mitigation strategies that is for its ecological, physical, and for the socio-economic impacts.

To formulate the effective management plans for a longer time for the reefs post-installation operations, these assessments are basically used. There are few questions that can be asked and can be answered before making any commitment in terms of planning. These questions are:

a. Is the concept of the reef realistic?

This question basically aims at evaluating, only if the construction of an artificial reef is valid and is feasible in a particular area.

b. How will the new reef and the natural ecosystem interact?

Before constructing the artificial reefs, it is important to understand the influence of an artificial reef upon the seafloor. That is in terms of how the natural habitat can be modified and how the new substrates may affect the ecological process. This is important for the long-term operation of the reef.

It is important to evaluate some of the situation like local, social, and economic. It is also important to involve the potential users of the artificial reefs in order to consider their opinion on the project.

Once the above questions are being answered, there are some more specific objective that should be defined. For example, increasing the income of the local fisheries up to some certain level. It is also necessary to calculate the investment as well as the expected ecological and socio-economic returns.

c. What are the potential alternatives to reefs construction which may produce similar environmental, social, and economic outcomes? What can be the situation if the reefs are not built?

These questions will also help in understanding whether the choice of constructing an artificial reef can be more effective than any other solutions.

Once a preliminary cost analysis has been performed and specific objectives have been established, after that the reef site can be definitively be identified and artificial reefs can be designed. It is important to take into account and the physical features of the proposed site that includes water depth, sedimentation of grain size and sediment cover, wave size and direction, etc., before selecting the reef site.

There other factors that are needed to kept in mind while selecting the reef site are, life history of the target species that includes distribution, reproduction, and feeding, and to note the specific site location for the purpose of the reef. With this, it is also important to note the type of material which must comply with the international and national legislations in order to avoid the dumping and pollution in the marine system or environment.

The artificial reef should be designed by the experts in the field. The designed should be based on the ecological and technical specification that are relating to its purposes and settings. The structural integrity of the reef unit and the subsequent group that is to be created should be given equal weightage in order to anticipate in their ecological effectiveness. When a reef is unstable, then this will ultimately fail to meet its ecological objectives.

In order to obtain their authorization, it is important to have the operational long-term management plan that must be submitted to the competent national or local authorities. These authorities are responsible for maintaining and monitoring the artificial reefs.

7.6.2. Construction phase

Construction phase includes all the activities that are related to the construction of the artificial reef structure and their deployment in the water. Generally, it is important to pay attention towards identifying the construction area and for granting access to a suitable marine staging.

It is also important to identify the site that may be suitable for that reef material storage, moving, and loading on the deployment infrastructure.

The selected area should have the easy access from both water and land. In order to facilitate the transportation of the reef material it is important to select the area which fulfils all the criteria. To avoid the risk, there are some safety parameters that should be established and should be signalled by the buoys around the deployment site to leisure the boats and the drivers during construction.

In order to avoid the risk to leisure the boat, drivers, etc., safety parameters should be established and signalled by the buoys at the development site during construction. The activity of transporting reef material from the staging area to the deployment area or site can restrict the manoeuvrability and it also represent the potential hazard navigations.

Figure 7.4. Reefs for catching a greater number of fishes.

Source: Image by Pintrest.

Therefore, it may be important to notify the relevant marine authorities. These authorities are said to have the jurisdiction authorities for the reef installation site of the proposed operational time table.

One of the main important problem that arises during deployment of artificial reefs. It is very important to place the reef structure at the place which is already preplanned.

Therefore, differential global positioning system (DGPS) is used to identify each site using this system and then marking those areas with the temporary marker buoys. After completion of deployment operation, scuba driver is supposed to check through direct observation, the correct position of the reef and its orientation of the structure.

They should also include the minimum safe navigation depth on the lowest astrological tide (LAT) known as LAT. Scuba divers uses surface deployed remote cameras or indirect surveys with the side scan sonar or multi beam echo sounder (MBES).

National or regional hydrographs authorities should be provided the final location of the reef unit which includes the boundary of the reef complex and safe navigation depth, so that they can update the reef location on the navigational map in order to bring them in notice to mariners.

7.6.3. Post-construction phase

After the construction of the artificial reef, the next question arises which is related to the reef management and uses in a sustainable way. These questions basically aim in identifying management options in order to optimize the benefits of the reef to all the users and to reduce the conflicts that arises between them.

But this aspect cannot work in the case where there are private entities. This means that this aspect does not work when an artificial reef is being constructed by the private entities, because they are considered to have the ownership on that reef site.

Therefore, this fundamental is applied only in the case where artificial reefs are being sponsored by the national, regional or local authorities that is available for the local communities. The other most important step is to determine the effectiveness of the artificial reefs.

The development of the scientifically valid monitoring programs is important with the establishment of the "success criteria." One of the most important criteria that can be used to enable the project sponsors to gain the evidence concerning the ecological and socio-economic performance of the reef which is in relation to the expected objectives is Monitoring.

7.7. REASONS TO BUILD ARTIFICIAL CORAL REEFS

Figure 7.5. *Types of coral reefs.*

Source: Image by Medium.

7.7.1. Block passage

The first artificial reefs were built with the help of the Persians to block passages for boats in ancient times.

7.7.2. Increasing fish yield

The next reason with respect to the artificial coral reefs was to elevate the amount of fish in a specific region and with it the yield for fisheries. On the other hand, this sounds like a great plan at first, but it might also have negative impacts. Read more about this topic in the pros and cons section mentioned below.

7.7.3. Conservation reasons

In the present interval of time, this is one of the main reasons in order to build the artificial reefs. Several numbers of reefs have been damaged with the help of human (or naturally as well) effects and the researchers agree that is a large danger to the oceans and humankind. Coral reef restoration of any kind is a very important tool in order to remedy these kinds of negative effects.

The methodologies in order to consider are either coral transplantation (growing corals and bringing them back to reefs where coral had died off) or the artificial reefs. In present interval of time, a common practice is an integrated approach of artificial reefs and coral transplantation to the structures.

7.7.4. Erosion

Artificial reefs can provide assistance in order to avoid the coastal erosion with the help of either forcing waves to break further away from the beach or to hold the sediment on the shallow regions.

7.7.5. Surfing

Artificial reefs built to break the waves at a particular point that helps in allowing more days when surfing is possible or produces the perfect waves.

7.7.6. Scuba diving

People love to see a lot of life in condensed or compact for in small region which is generally seen in artificial reefs. In this way, several numbers of scuba divers love to dive in any kind of artificial reef. There are several kinds of artificial reefs that are mentioned in later topics.

7.8. DIFFERENT TYPES OF ARTIFICIAL REEFS

Even underwater structures that are made by man-power, which permits coral to prosper on it and a reef in order to form around it is an artificial reef. The structure provides a hard surface where algae and the coral reef become attached.

Coral starts to link and with its small life, it will determine its way to live between the structures. The small creatures will attract the bigger fish and in the end, the artificial coral reef will be packed with life.

7.8.1. Wrecks

Trains, bicycles, ships, oil rigs, cars, and other wrecks operate as artificial reefs. In addition to this, in the recent interval of time, ships that sank in the wars or storms, several numbers of them were sunk on purpose in order to build the artificial reefs and an attraction for the divers.

These wrecks stripped from any harming structures on the inside and the holes are created which permits easier entries with respect to the divers. As well as wrecks, completes structures, for example, jetties, every so often operate as artificial reefs. On the other hand, they are still in use. These can be great for fish in order to hide and in the context of scuba drivers, it is easy to take fascinating photos underwater.

7.9. ARTIFICIAL STRUCTURES BUILT USING DIFFERENT MATERIALS

These kinds of artificial coral reefs can be made out of various kinds of materials such as rock, metal, concrete, and things like that. They build in different manner which relies on the purpose such as increment in the fish yield, safeguarding the shores, breaking waves, restoring the reefs or entertaining the scuba divers.

Most of the structures that scuba divers address with, are built in such a way that they offer a place to hide for marine life and facilitate the attachment of the corals. Every so often, even the sculptures have been placed in specific areas in order to create the underwater museums.

Studies are still going on, to find the best material or the combination of the materials in order to build the suitable or perfect artificial reefs.

Generally, rocks are a natural ground to which coral can become attached and prosper. In this way, this is an obvious selection providing it is the right kind of stone (for instance, limestone). No one cannot just throw any kind of rock in the ocean and claim that he or she have built an artificial reef.

On the other hand, wood does not last as long as other materials would, it is natural in nature, and with respect to particular kind of region it already forms part of the underwater world. Several numbers of artificial reefs are the blends of the above mentioned.

7.10. BIO-ROCK

This is unique kind of technique where a conductive frame that operates on electricity (e.g., wire mesh) is installed as an artificial reef and then a low voltage is applied on it. This produces a reaction that permits the minerals naturally found in the seawater in order to grow on the structure.

The material is very identical to the coral reef and as long as the current is applied it will endure to grow. This causes the structure to become stronger and more rigid over the interval of time as well. The electrical fields as well as the protection is provided by the structure that helps in attracting the extensive range of the marine life.

Corals that are attached to the Bio-rock has been shown to withstand more dramatic conditions as compared to the coral in the natural reefs around them.

7.11. RUBBLE OR CONSTRUCTION DEBRIS

On the other hand, some of these materials can provide a great basis for invertebrates to become attached, some can do more harm as there is possibility that material could be toxic. There is also a danger that the debris would just be dumped in the ocean as it is a much easier way to get rid of it without providing much benefit.

More particularly, tires that have been found in order to make fairly bad artificial reefs and in several numbers of places these reefs have had to be removed again.

7.12. PROS AND CONS OF ARTIFICIAL REEFS

7.12.1. Cons of artificial reefs

Never the less, there are possible negative impacts of the artificial reef structures as well. The development of the artificial reefs in order to provide the additional habitat for marine life and restore lost coral reefs has been a trial and error process.

Several numbers of projects have failed and some of them are even ended up in doing more harm than good.

1. **The material might become toxic:** Several numbers of material are consisting of rubber and the metal will degrade or corrode that release toxins substances in to the water.
2. **Tires didn't make great artificial reefs:** Enormous number of old tires were sunk to the ocean floor in several numbers of nations. This was seen as a great way to get rid of the waste, on the other hand, some building something great under the water. Never the less, this did not work out as it was planned.

The tires do not offer a good basis that the invertebrates, for example, coral can become attached. In addition to this, through the course of storms or due to the attachments to the ocean floor wear off tires may get loose.

They might get washed up on the natural coral reefs that destroys the complete structures on their way to the shore. In several numbers of regions, the tires have now been removed again.

3. **Overfishing instead of increasing biomass:** In addition to the above mentioned, there is another concern with respect to the artificial reefs is that, they do not actually elevate the amount of fish, but also, concentrates them around a particular region.

The higher concentration of fish makes it much easier for the fishers in order to catch them. This might even encourage with the increment in the overfishing issues, in spite of helping the marine life.

In order to prevent this artificial reef, maybe all reefs, should be safeguarded and fishing only allowed at a particular distance from the structure.

4. **The artificial reefs might be different to natural ones:** Researchers have compared the marine life around two shipwrecks (out of which, one is five years old and the other is 105 years old) with natural reefs in the same region. They found that the "artificial as well as the natural reefs strongly differ in their trophic structure, both in their multivariate composition and in the biomass of the most guilds."

In simpler words, the kind of the marine life as well as the amount also fluctuates between the two. This means that the ship wrecks might not be appropriate as a replacement for the lost natural reefs and the ecosystem around it is different as compared to the other natural reefs. This might be true for the other artificial reefs as well.

There are also concerns that the particular kind of artificial coral reef structures might actually attract the fish away from the natural coral reefs and producing an imbalance in the natural habitat.

7.12.2. Pros of artificial reefs

While it has been observed that there is a lot of research that are going as to how to build the right artificial reefs in order to improve the state of the oceans. Also, there are no studies done that can prove an actual positive influence on larger regions.

Yet it is considered as one of the best approaches to helpful in recovering the oceans from reef loss and the subsequent decrease in biomass.

1. **Coral reef restoration**

Artificial reefs play a very important role in the restoration of coral reefs. These coral reefs have been destroyed by the impact of human or natural disasters such as storms or El Niño.

One should make use of right material for the structures which is like a natural basis for coral growth and transplanting coral to it helps in permitting the restoration of lost reefs hopefully faster than they would regrow in natural way.

2. **Strong against erosion**

It has been observed that shallow areas that are consisting of sand and gravel might be washed off in strong currents and big waves. The artificial reefs can be built in order to withstand such kind of impacts. Thus, such kind of structures can help in making shores more stable against erosion.

3. **They attract tourists**

The artificial reef helps in attracting more tourists to a destination, from surfers to snorkelers to scuba divers. The economic value will help in increasing the interest in shielding the ocean even more.

So even taking into accounts all the cons, artificial reefs are a good way of trying to enhance the underwater ecosystem and there is a need of a lot more research in order to find the best way to do so. While everyone is in the process of finding some good options of restoring habitats for marine life, artificial reefs are considered as a good start as long as they are not noxious and are attached to the floor of the ocean properly.

7.13. FISHERIES MANAGEMENT ISSUES

The techniques that are relevant to the management of the fisheries on the artificial structure is not very different but kind of similar to general fisheries management tools. But these artificial structures of reefs raise some specific issues that are related to the availability of the data.

Some of the issues are related to the alleged effects on the resource concentration and generation. Their potential role is to conflict the resolution, resource allocation, coastal protection and for the development and their cost-effectiveness.

7.13.1. Information requirements

It is important to have the proper management of the reefs and FADs. As they imply the availability of the biological, economic, and sociological data. Most of the reviews are on this issue which recognize the situation which is presently totally inadequate when it is considering the investment that is involved (Bohnsack and Sutherland, 1985, ASEAN/UNDP/FAO, 1988).

There are numerous biological problems (cf. Polovina). They are said to be involved or related to the several issues that basically includes the evaluation of the biomass and annual production. There are some other issues like optimization of the energy transfer from the primary production of the fishable biomass, separation that takes place between the new or additional and attracted production, and issues that are related to the interaction that takes place between adjacent reefs.

The major difficulties lie in forecasting and during assessment of the biological impact that is in relation to the site, reef types, and fishing methods.

Huntsman in the year 1981 indicated that by chance all the reef fishes are herbivores and small carnivores. It is a fact that consumers basically prefer large carnivores which contain low natural mortality and usually have slow growth.

They prefer those fishes that implies high age at first capture and low fishing mortality. Therefore, in turn implies that the reefs are useful for sports or for recreation of non-intensive fisheries rather than turning towards the commercial fisheries.During Research there are several problems arising in marine fisheries. These problems are due to multispecies that are interacting and determinism of recruitment that is needed to be urgently addresses. It does not matter how difficult it can be to resolve.

There is a need for closer attention that is to be given to the link that is between the local environment, primary production and fishable resource. These all needed I order to understand and forecast the abilities that must be developed.

The social and economic considerations are far from the trivial. They are related to the identification that is related to the fishery constituency, Benefits, and their distribution are for understanding and forecasting, the rate of understanding of the reef, the appropriate collection and identification of sociological data, demographic data and attitudinal data.

There are some major difficulties that lies in the undertaking of the baseline studies and for the classification of the fisheries relationship that is between the reef fishery and the ambient fishery. There are some legal implications that are still largely undebated.

7.13.2. Concentration of resources

Both the types of artificial structures can operate as attracting devices, which in turn increases the catchability of otherwise more dispersed resources. Increase in the fishing efficiency and economic returns are considered to be the consequence and therefore, there is a risk of aggravating overfishing if in case effort is not regulated.

In short term, profitability for the fishermen may increase because of the reduction in operating costs. This is done by localizing the fish concentrations in much efficient way and reducing the searching time relative to fishing time.This can provide incentives for a number of fishermen in order to enter the fishery if the effort in the area is not controlled, this in turn can aggravate the overfishing problem. The increase in the catchability and in effort will result in a general increase in the fishing mortality and also, a corresponding reduction of the stock size, catch rates as well as probably total catches.

Therefore, it is evident that the artificial structures cannot be considered as a solution to the overfishing problem which is growing (Polovina, 1989). Aprieto (1988) also observed this and further noted that large quantities of juvenile yellowfin as well as skipjack of 16–20 cm are captured on the payaos in the Philippines.

There are some reactions from industry which are already apparent, despite the fact that the impact on the surrounding tuna fisheries is not yet clear.

There is one more effect of resources concentration and that is by reducing the costs of fishing, especially when the construction as well as deployment costs are subsidized, the economic overfishing problem may alleviate by the artificial structure but only temporarily, if in case, there is no control on effort.

7.13.3. Resources enhancement

Considerably, it can be said that in the reef-related bibliography, most of the time, management is taken in its broad sense which includes resource enhancement as well. Most probably, artificial reefs can produce, some additional biomass (that can be done at least locally) by:

- Trapping biological productivity which would otherwise be transported elsewhere (like in the Adriatic Sea);
- Improving pre-recruitment survival through the range of shelters offered to larvae and juveniles, and; and
- Accelerating natural reefs regeneration.

If they are installed in non-reefs area, then there are chances that they can allow the growth of non-traditional species which in turn leads to biological diversification. Good scientific assessments are very rare. In short term, this seems to result in higher catches, species, and better returns.

However, if the effort is not properly controlled, then this will provide the incentives for a number of fishermen to enter the fishery, leading to an aggravation of growth overfishing and to the dissipation of additional rent which is created by the reef.

The catches can be higher and also, the employment which can be said to be a consequence, but the catch rates as well as profitability would return to original and low equilibrium levels.

There are artificial reefs installed in the vicinity of natural reefs areas and the problems of them are different. According to the arguments presented by Polovina in the year 1989, if the decrease in resources is because of the recruitment overfishing, the artificial reefs are a costly as well as inadequate response since, there cannot be a shortage of adequate habitat wen the stock of spawned has been reduced by doing fishing to less than half of its virgin size.

Considering some of the cases of the artificial reefs, it can be said that they have been established as protected marine enhancement areas, where the fishing if prohibited like in Malaysia and Singapore.

It is obvious that in the overfished coastal zones and areas, it is important to have surveillance for this approach which is much costly, considering the reference to the argument provided by Polovina. This argument is said to be justified only when there is decrease in the resources which will be due to habitat destruction.

In that case, stopping habitat degradation becomes a priority. However, in the case of FADs, there have fears that the concentration of juveniles, along with the large predators, can result in increased natural mortality by predation as well as cannibalism and therefore, reducing the stock potential (Aprieto, 1988).

7.13.4. Conflicts for space and resources

Paradoxically, reefs as well as FADs are also considered as a limited resource (if there is limitation of space which is available for them) and they can further create an allocation problem, instead of serving like a means for reducing conflict.

There is an inevitable consequence of the concentration of fish and that is the concentration of fishermen especially around fewer fishing spots. This can result in gear saturation as well as competition, while lowering the returns and also, in enhanced conflict.

For instance, Aprieto, in the year 1988, showed that payao fishermen in the Philippines agreed to a minimum distance equals to 7 miles between FADs. Reefs as well as FADs can also create conflicts between commercial or recreational fishermen, between divers, anglers, and also, the commercial fishermen, or, between sport fishing and photography, between fishing and navigation.

These conflicts can said to be difficult to resolve, but it might be desirable to create different reefs so that the conflicts can be diluted in between incompatible types of uses instead of trying to reconcile these conflicting uses within a single large artificial reef.

The coastal space can be occupied effectively by using the artificial structures in order to impede the intrusion of trawlers into coastal areas. An example can be seen in Thailand (Sungthong, 1988).

This can prove to be an effective way in order to decrease conflicts that are created in between artisanal and trawl fisheries. The aim can be focused on the partial or total protection of the reef's resources and so, the artificial reefs can be used for creating sanctuaries. As an example of this can be seen in Malaysia (Wong, 1988) and Singapore (Hsu et al., 1988).

7.13.5. Cost-effectiveness

As there may be some economic gain of artificial structure. But there are number of conditions that have to be fulfilled in order to ensure the net economic and social benefits.

If the artificial structures are considered as the most cost-effective method then the net pains will be released for achieving the intended purpose. Actually, in practice there has been very little efforts made in order to try the real cost-effectiveness of these devices.

There are some parameters that are generally supported by the state through subsidies and that is the cost of the initial investment which a relevant parameter in this respect. It totally depends on the on the improved rates that is used for in decision making that is made on the allocation of the rent that is eventually created between the state by the payment of a fishing rights and the users.

While dealing with the growth or recruitment overfishing, it is obvious that the artificial structures can be cost effective. This is particularly from the problem of excess fishing capacity and with the open access that are not faced.

7.13.6. Allocation of user-rights

Artificial structures can be utilized to openly allocate the marine space as well as the resources to the identified social strata of the fishery sector for their exclusive use. This can be even the prime reason of the artificial reefs that are installed in a natural reef region.

If the province of the space by the reefs and FAD (Fish aggregating (or aggregation) device) is very efficient enough in order to conduct the efforts of fishing of trawlers out of the costal nursery areas, there should be a general increment in the biomass because of the decrement in the mortality and increase in age at the first capture.

Never the less, the real effect is required to be carefully evaluated in each of the case, the example of these kinds of total as well as temporary trawl bans in the coastal areas under severe overfishing have already proved to the outcomes in meaningfully larger biomasses (Garcia, 1986; Dwiponggo, 1990).

The example in the Cyprus shows that the recovery can be extremely fast (near about 100% in the last two years), but also that the response of the small-scale fishery can be just as quick in expanding the effort and dissipating the rent if the effort is not regulated.When the coastal region is overfished and the fishing spots are much crowded, artificial reefs as well as the fish aggregating devices can provide the alternative to fishing grounds. In several regions of the United States of America, for example, anglers, and sport fishermen necessitate the reefs for their own purposes. Here, the artificial structure can provide assistance in order to reduce, (at least temporarily), conflicts for space with the help of developing more space for fishing. The long-run impact, never the less, will rely on the overall trend in fishing effort.

Generally speaking, it may be advisable to well think out explicitly the conflict and allocation problems at an early stage of reef deployment planning in order to prevent it from becoming only an additional subject for the conflict.The problem of the property rights is more critical for the artificial reefs are practically permanent structures, as compared for fishing aggregating devices (FADS) which can be easily removed.

Typically, property rights would provide security (tenures), decrement in uncertainty, increment in the existing value of the future production, improvement in the local support with respect to the collection of data, research, and resource management.

7.13.7. Habitat protection and rehabilitation

Artificial reefs can take participation with the environmental migration with the help of decreasing the effect of the trawling on the fragile habitats as well as nurses. And also, they are used in order to recompense for the destroyed coastal habitat or for natural reef generation (e.g., in the Philippines and Indonesia).Artificial reefs may also provide assistance in order to enhance the environment on the land with the help of offering the chances to dump a large number of tyres, car wrecks, etc.This opportunity should obviously not hide the danger of application of reefs as a pretext to dumping unqualified that may badly harm the marine environment. Artificial reefs can also serve in order to safeguard the coastlines from the erosion and from the effect of the storms (e.g., Japan).

7.13.8. Development

These structures can also be used in order to develop new fishing spots close to villages for commercial, subsistence or recreational purpose.

For example, the payao system in the Philippines has taken into the direction to the development of a major tuna fishery that is shared among an artisanal fishery for mid-water resources and the industrial purse seining for the surface resources (Apricto, 1988).

The attraction, improvement, and changes of resources, as an outcome from the artificial structures could, in several numbers of cases, lead to the economic gains and the enhanced performance of the fishery sector. Whether these kinds of impacts are really obtainable depends on how carefully the structures have been planned.

Whether the enhancement will be sustainable relies on a proper reef and fisheries management because unsettled or unanswered allocation problems and as an outcome, its conflicts will surely decrease the economic performance.

The development of the rural regions and the conservation of the socio-economic activity in remote coastal or seashore regions are always been a prime concern, but not exclusively, in the context of emerging nations (e.g., Japan).

The enormous and sustained introduction of reefs takes into the direction to develop a meaningful economic activity in the coastal region, as an outcome from the construction of the reef, its maintenance, and the exploitation.This impact and the political aid it produces is apparently is enough in order to justify the large subsidized investments in the nations such as Japan, as an alternative of the apparent absence of proper economic analysis.

7.14. CONCLUSION

Figure 7.6. Fish for artificial reef.

Source: Image by Wikipedia.

This chapter tell the methods that is used for developing artificial reefs. As before planning to set the reefs, it is very important to see all the criteria that are important to be kept in mind. As these factors are very important before choosing any area for the reefs.

Choosing areas that are particularly important before setting up the reef is very important. These artificial reefs were planned as a fishery management tool, which were expected to increase the fishing yields and in order to reduce the conflicts which exists among various fisheries sectors. This fisheries management was also used as a protection device considering the coastal sensitive habitat. It is very important to work according to the legislation rules and regulation. Planning is also an important aspect of artificial reefs. Setting up reefs is not an easy task. So, there are many factors that should be kept in mind before planning it.

REFERENCES

1. Asmfc.org. (n.d.). *Artificial Reefs-Atlantic States Marine Fisheries Commission*. [online] Available at: http://www.asmfc.org/habitat/artificial-reefs (accessed on 8 March 2020).
2. Fabi, G., Bortone, S., Brandini, F., & Otake, S., (2011). *Artificial Reefs in Fisheries Management*. [online] ResaerchGate. Available at: https://www.researchgate.net/publication/313876631_Artificial_Reefs_in_Fisheries_Management (accessed on 8 March 2020).
3. Fabi, G., Scarcella, G., Spagnolo, A., Bortone, S., Charbonnel, E., Goutayer, J., Haddad, N., Lök, A., & Trommelen, M., (2015). *Practical Guidelines for the Use of Artificial Reefs in the Mediterranean and the Black Sea*. [eBook] Available at: http://www.fao.org/3/a-i4879e.pdf (accessed on 8 March 2020).
4. Florida Fish and Wildlife Conservation Commission, (n.d.). *Artificial Reefs*. [online] Available at: https://myfwc.com/fishing/saltwater/artificial-reefs/ (accessed on 8 March 2020).
5. Garcia, S., (n.d.). *Artificial Reefs and Fish Aggregating Devices in Southeast Asian Fisheries: Management Issues*. [eBook] Available at: https://horizon.documentation.ird.fr/exl-doc/pleins_textes/pleins_textes_6/b_fdi_49–50/010013396.pdf (accessed on 8 March 2020).
6. More Fun Diving. (n.d.). *Artificial Coral Reefs-Types, Reasons, Pros, and Cons-More Fun Diving*. [online] Available at: https://morefundiving.com/artificial-coral-reefs/ (accessed on 8 March 2020).
7. Pipitone, C., & D'Anna, G., (2014). *Marine Managed Areas and Fisheries*. [online] Sciencedirect.com. Available at: https://www.sciencedirect.com/topics/agricultural-and-biological-sciences/artificial-reefs (accessed on 8 March 2020).
8. Polovina, J., (1989). *Artificial Reefs: Nothing More Than Benthic Fish Aggregators*. [eBook] Available at: http://calcofi.org/~calcofi/publications/calcofireports/v30/Vol_30_Polovina.pdf (accessed on 8 March 2020).
9. ScienceDaily, (n.d.). *Artificial Reef*. [online] Available at: https://www.sciencedaily.com/terms/artificial_reef.htm (accessed on 8 March 2020).
10. Watanuki, N., & Gonzales, B., (2006). The potential of artificial reefs as fisheries management tools in developing countries. [online] Available at: https://www.researchgate.net/publication/233680292_The_potential_of_artificial_reefs_as_fisheries_management_tools_in_developing_countries (accessed on 8 March 2020).

8

Future of Market-Based Fisheries

CONTENTS

The chapter of the future of market-based fisheries talks about the basic significance of the trends of the fisheries in the present interval of time. This chapter also explains the current trends in seafood trends. This chapter also addresses the present as well as future markets for fish and fish products from small scale fisheries.

This chapter also provides highlights on the Small-Scale Fisheries (SSF) production such as capture fisheries production, culture fisheries production, and composition and value of SSF production.

This chapter also consists of the future of fisheries management which includes the size of the fish, small fish in a big pond, the race for space, fisheries in changing climate. This chapter also addresses the future of the industry of fisheries.

8.1. INTRODUCTION

Fishery industries have been taken for a several numbers of reasons over the millennia, and instances of overfishing or other influences that are unacceptable can be found in several numbers of various fisheries, for example, sport and market, subsistence (Jackson et al., 2001, McPhee et al., 2002, Mullon et al., 2005).

Never the less, in spite of commonly held views, overfishing, and the other influence on the fisheries have not arisen simply as an outcome of industrialization of fishing in the recent past (Jackson et al., 2001).

In spite of these issues, the demand for seafood has increased with a very quick pace, placing hitherto unknown demands on the resources and, it could be debated, has developed the requirement for the new tools in order to make sure that sustainability is the dominant factor that helps in guiding the future utility of the resources of seafood.

Conventional methodologies with respect to the fisheries management have depended on developing a series of rules in order restrict the production. These kinds of rules may be approved by the government or administered with the help of non-government organizations, for example, fishing associations, but irrespective of this fact, the intended effect is to regulate the supply of seafood products.

The demand side has largely remained unaffected, at least as far as sustainability is concerned, expect in an indirect sense when the impact of the scarcity and the simple market abundance or absence effect on the price may have an impact on the demand.

With respect to the active influence on demand, sellers of the seafood have always sought to develop the demand with the help of advertising some of the characteristics of the seafood in order to increase the sales.

Figure 8.1. Basic significance of the fisheries.

Source: Image by Pixabay.

Attempts to directly impacts the demand for the purposes of impacting the sustainability are a comparatively phenomenon. As will be set out below, the rationale for these kinds of interventions is based on the several numbers of the instances of the failure (total or partial) of Supply-side catch constraints.

This does not mean that these kinds of restrictions are valueless, but that demand-side controls can provide the additional solutions. Used together, a more effective system can result.

Seafood is one of the most highly traded commodities all across the globe. Supply chains may be complicated and long, extending all across the hemispheres, and in this way, supply, and demand may not be obviously linked.

Never the less, as mentioned below, the foundation of the chains of custody that provides explicit association between the supply and demand is important and feasible irrespective of the nature of the trade routes.

8.2. TRENDS IN SEAFOOD TRADE

The Food and the Agriculture Organization (FAO) has documented the rapid expansion of the international trade in the seafood over the time period of the nearly two decades. The value of the global national exports has increased from 7 billion US dollar in 1976 (Josupeit & Franz, 2003) to 58 billion US dollars in 2002 (Lem, 2004).

Net export revenues with respect to the developing nations grew from less as compared to the 4 billion US dollar to 15 billion US dollar over the same interval of time.

Never the less, some nations have always been a meaningful trade of the seafood and the seafood products, the increment in the past several numbers of years have been because of the rapid expansion of catches in the waters of developing nations and the export of the seafood to developed nations.

Near about 22 million tonnes of seafood entered the world trade in year 1996, which depicts an increment of 8 million tonnes since the time period of 1976 (Deere, 2000).

Dommen (1999) predicted that near about 40% of the seafood all across the world now enters into the international trade (as compared to 10% of meat; Delgado et al., 2003) and the proportion that participated with the help of developing nations had been increased from the value of 43% to 51%, over the interval of time from 1992 to 2001 (FAO, 2007).

Most of the trade took place in between the emerged nations and other emerged nations, or between emerging nations and emerged nations (Deere, 2000).

In the time period of 2001, most of the seafood (74%) was exported to Japan (23%), the United States (17%) and the European Union (EU) (34%) (Vannuccini, 2002). The trade is done between the developing nations which is dominated (by weight) with the help of fishmeal and fish oil for the use in aquaculture.

Aquaculture production has grown with a rapid rate from 15% of the seafood all across the world in year 1990 to the value of 29% in the year 2001. The increment in bulk of this production has took place in the People's Republic of China, and the production is dominated with the help of freshwater fish (Subasinghe et al., 2001).

On the other hand, the industry of aquaculture is one of the few primary industries in several numbers of nations which is expanding (and is the main hope for fulfilling the increased demand for seafood), this increment has not been without a significant environmental influence in some regions (Subasinghe et al., 2001).

Co-incident with the increment in the trade of the seafood has been an alarming deterioration in the state of several numbers of fisheries all across the globe. Over the time period of several numbers of decades, the FAO (2005a) has documented that a gradual increment has been shown in the number of fisheries.

That are listed as over fished (in present interval of time, it is 17%), and there is concern that several numbers of those that are presently listed as fully fished (which is near about 52%) are merely in state of transition to an overfished status.

Figure 8.2. Trends in seafood trade.

Source: Image by Pikrepo.

As the global catches have been increased (in present interval of time, it is 82 million tonnes), so too has the area of waters fished. In the year 1980, the high seas (waters outside of the 200 nautical miles Exclusive Economic Zones (EEZ) is claimed with the help of maritime nations) provided 5% of the world catch.

This has been increased to 11% in the year 1990 (Deere, 2000), in this way, developing a variety of management and enforcement problems that endures to produce concerns, more specifically, with respect to regulating trade.

The loss of food fish from decreasing fisheries in only one of the factors to have a concern about it. Fishing has its own influences or effects and there is a global concern about the effects of fishing on groups of species, for example, sharks, turtles, and seabirds, along with a more general ecosystem modification (FAO, 2001).

In the context of coastal regions, fishing pressure adds to an extensive suite of pressures on waterways (estuaries and rivers), for example, dams, urban development, pollution, and wetland drainage. Certainly, the coastal development is proceeding with such a speed in several numbers of nations that the future inshore fish stocks is bleak (FAO, 2007).

8.3. PRESENT AND FUTURE MARKETS FOR FISH AND FISH PRODUCTS FROM SMALL-SCALE FISHERIES

The sector of marine fisheries has contributed near about 1.62% to the GDP (Gross Domestic Product) of India, 1.5% to the GDP of Malaysia and near about 1.9% to the GDP of Thailand.

With respect to these three nations, the fishery sector has provided fish as a source of protein as well as opportunities of employment for hundreds or thousands of people and are very important earners of foreign exchange.

With respect to India, near about 1.284 million active fishers are involved in the marine capture fisheries alone. Talking about Malaysia and Thailand, near about 82,630 and 1, 61,670 fishers, respectively, are contributed in marine capture fisheries.

In these three nations, several kinds of fishing craft and gear are ranging from conventional non-motorized boats to the modern factory vessels are being used in order to harvest the fishery resources. The numbers of fishing vessels in India is near about 2, 80,491. With respect to Malaysia and Thailand, they have near about 30,751 and 1, 18,436 fishing vessels, respectively.

Most of the fishing crafts are ranging from seventeen feet to twenty-eight feet (which is near about 5 meters to 8.5 meters) in length and are driven with the help of inboard as well as outboard engines, along with the help of sails in case of conventional non-motorized boats.

Several numbers of various kinds of gears, for example. Loglines, trawls, hand lines, poles, and lines, gillnets, push nets, and purse seines are being taken into use.The majority of the fishing vessels functions in the coastal waters, which are entirely and every so often overexploited.

The definition of the SSF in these three nations depicts much diversity. SSF in Thailand is defined, with respect to the length of the boat, types of gears used for fishing, horsepower, and labor employed.

SSF are defined as fisheries, which use conventional gears of fishing and fishing boats having a maximum length of near about ten meters and an engine which produces maximum horsepower of 30 hp, functions in the vicinity of the home base of the owner of the vessel and close to the coast.

In several regions of Thailand, SSF are further defined with the help of the low incomes, being a subsistence-level economic activity and by the matter of fact, most of the labor is provided with the help of family members and the relatives. Coastal aquaculture of small scale also falls under the category of SSF.

Figure 8.3. Present and future markets of fish and fish production from small scale fisheries.

Source: Image by Pxfuel.

Talking about Malaysia, the term "SSF" are limited to the fisheries with the help of non-motorized and outboard-powered fishing boats and to all inland capture fisheries. In the context of India, more than 66% of the entire fishing and fish processing industries are categorized as a small-scale with the help of government.

8.4. SMALL-SCALE FISHERIES PRODUCTION

8.4.1. Capture fisheries production

In Malaysia, the total marine fish landings have been increased by the value of 0.86% from one thousand two hundred and seventy-two tones in the year 2002 to near about 1283 million tones in the year 2003.

As mentioned in the table given below, in Thailand and India, the total marine catch has been decreased by the value of 0.87% and 2.05%, respectively, over the same interval of time. In all three nations, the bulk of the production came from the inshore fisheries, which accounted for more than 80% of the total marine fish production.

In Malaysia, the conventional gears that are used for fishing are operated with the help of near about 23,565 fishing vessels produced 277,847 tones of fish, which is valued at near about 0.25 billion USD. On the other hand, gears such as hooks and lines produced near about five hundred and eighty-six tonnes of fish, which values near about 0.78 million USD.

In Malaysia, the contribution of the SSF to the marine capture fisheries production has been valued near about 6.8%, 70% in the case of India, and 7.2% in the case of Thailand. The contribution of SSF to the marine capture fisheries production and its value is shown in Table 8.1.

Table 8.1. Marine capture fisheries landings in india, malaysia, and thailand and the contribution of Small-Scale Fisheries (SSF) in 2003

Country	Total landings (tonnes)	Share of SSF (%)	Quantity of SSF landings (tonnes)	Approximate value (US$ million)
India	2,977,965	70.00	2 084 575	1 373.66
Malaysia	1,283,256	6.80	87 261	75.93
Thailand	2,620,582	7.22	189 387	229.78

Source: table by FAO.org.

The inland capture fisheries production in India, as well as Malaysia, has shown an elevated trend through the course of 2003, on the other hand, the inland capture fisheries production marginally decreased in Thailand.

In the context of Malaysia and India, the production showed an increment of near about 378 tones (10.96%) and 15,960 tones (2.51%), respectively. On the other hand, in Thailand, it has been reduced by the value of 1800 tones (0.91%).

The entire production of the freshwater fish from capture fisheries in the three nations is contributed by SSF exploiting rivers, lakes, and reservoir as shown in Table 8.2.

Table 8.2. Inland Capture fisheries production in India, Malaysia, and Thailand and the contribution of Small-Scale Fisheries (SSF) in 2003

Country	SSF landings (tonnes)	Share of SSF (%)	Approximate value (US$ million)
India	651,719	100	292.90
Malaysia	3565	100	6.40
Thailand	198,400	100	151.46

Source: table by FAO.org.

8.4.2. Culture fisheries production

The industry of aquaculture in the nations studied have a large potential and can significantly took participation to their total fish requirements. All three nations have accomplished the progress in aquaculture production with the help of planning carefully and have adopted efficient or productive strategies for further growth.

In all three nations studied, aquaculture production has shown an elevated trend for the time period of last two years. It has been observed that the government of these three nations, with the help of their departments of fisheries, have laid emphasis on the increment in the outcome from the aquaculture farms, which is irrespective of their sizes, whether it is small-scale or a medium-scale.

Never the less, it is a challenging task to give a precise figure of the number of the aquaculture farms that are functioning in the three nations, the numbers of farms in Thailand has been approximated at near about thirty thousand.

In the context of India, the number of aquaculture farms is approximated near about ninety thousand, and with respect to Malaysia, the number of farms has been approximated at near about 10,000 to 15,000 farms.

With respect to all three nations, the aquaculture production for the year of 2003 endure to consist mainly of crustaceans, variety of finfish, and molluscs. The increment in the production through the course of 2002 to 2003 was near about 1.30% in India, 1.24% in Malaysia and 24.4% in Thailand.

In the time period of 2003, inland freshwater aquaculture accounted for near about 94.9% of the entire aquaculture production in India, about 31.8% in Malaysia, and near about 41% in Thailand.

Figure 8.4. An illustration of inland aquaculture.

Source: Image by Wikimedia commons.

In all three nations, the inland aquaculture production has been elevated in the year 2003 as compared with the year of 2002. In the context of India, it has been increased by the value of 1.5%.

The increment of the inland production in Malaysia and Thailand was near about 12.4% and 9.7%, respectively.

Talking about India, the complete production through inland aquaculture comes from the SSF sector. In the context of Malaysia, the contribution of the SSF with respect to the inland production is approximated to near about 10%.

On the other hand, in Thailand, it is near about 82.5% of all freshwater aquaculture farms are small-scale farms, which accounts for near about 66% of the entire freshwater aquaculture production.

In all three nations, the freshwater pond culture spearheaded production with the species such as catfish, common carp, climbing perch, snakehead, and freshwater prawn contributing to the bulk of the production. Freshwater eel, Tilapia, and barbs are also produced with the help of aquaculture in Malaysia and Thailand.

The inland aquaculture production and the contribution of the SSF as shown in Table 8.3.

Table 8.3. Inland aquaculture production in india, malaysia, and thailand and the contribution of Small-Scale Fisheries (SSF) in 2003

Country	Total landings (tonnes)	Share of SSF (%)	SSF landings (tonnes)	Approximate value (US$ million)
India	2,102,350	100.00	2,102,350	1,181.10
Malaysia	53,211	10.00	5 321	7.17
Thailand	361,125	68.60	247 732	217.79

Source: table by FAO.org.

In the year of 2003, the coastal and the brackish water aquaculture accounted for near about 5.1% of the total aquaculture production in India. On the other hand, the participation of this sector was as high as 68.2% in Malaysia and 59% in Thailand.

In the context of Thailand, the coastal and brackish water aquaculture production have been elevated in the year of 2003 as compared to the year of 2002. In Malaysia, the coastal and brackish water aquaculture has been elevated from the value of 48,375 tonnes in the year of 2002 to the value of 49,544 tonnes in the year of 2003, recording a growth of the 37.3%. Over the same interval of time, Indian production is decreased by the value of 1.5%.

In India, 90% of all brackish water shrimp farms have a size of 2 ha and another 5% of the farms have a size from 2 to 5 ha. Farms, that are having a size of 5 ha or less are categorized as a small-scale. These farms account for near about 95% of the coastal and brackish water aquaculture production of India.

In the context of Malaysia, only 5% of the coastal and brackish water aquaculture production produces from the small-scale farms. On the other hand, the participation of the small-scale farms to the coastal and aquaculture production in Thailand is approximated at 45%.

In the nations such as Malaysia and Thailand, coastal aquaculture and brackish water aquaculture production are having the concentration on the farming of marine finfish, shrimp, and crustaceans. On the other hand, in India, only giant brackish water shrimp, that is black tiger shrimp and white shrimp were cultured.

The production of the coastal and the brackish water aquaculture in India, Malaysia, and Thailand and the participation of the small-scale sectors are showed in Table 8.4.

Table 8.4. Coastal and brackish water inland aquaculture production in india, malaysia, and thailand and the contribution of Small-Scale Fisheries (SSF) in 2003

Country	Total landings (tonnes)	Share of SSF (%)	SSF landings (tonnes)	Approximate value (US$ million)
India	113,240	95.00	107,578	636.18
Malaysia	113,949	5.00	5,698	8.45
Thailand	454,275	45.00	204,423	983.65

Source: table by FAO.org.

8.4.3. Composition and value of small-scale fisheries production

In India, on the average, a tonne of marine fish produced with the help of marine capture fisheries is valued at near about 658.9 USD. On the other hand, a tonne of freshwater fish is valued at 449.4 USD.

Talking about Malaysia, the average value of the marine fish is 869.8 USD per tonne. On the other hand, inland freshwater fish is valued at around 179.5 USD per tonne.

In Thailand, the average value of marine fish is 1213.3 USD per tonne and 763.4 USD per tonne of freshwater fish. The SSF production in India, Malaysia, and Thailand, its source and its value are mentioned in the table 8.5.

Table 8.5. Small-scale fisheries production, value, and sources in India, Malaysia, and Thailand in 2003

Source/country		India	Malaysia	Thailand
Marine capture fisheries	Q	2,084,575	87,261	189,387
	V	1,373.66	75.93	229.78
	UV	658.96	870.15	1,213.28
Inland capture fisheries	Q	651,719	3,565	198,400
	V	292.90	6.4	151.46
	UV	449.43	1,795.23	763.41
Inland aquaculture	Q	2,102,350	5,321	247,732
	V	1,181.10	7.17	217.79
	UV	561.79	1,347.5	879.14
Coastal and brackish water aquaculture	Q	107,578	5,698	204,423
	V	636.18	8.45	983.65
	UV	5,913.66	1,482.98	4,811.84
Total	Q	4,946,222	101,845	839,942
	V	3,483.84	97.95	1,582.68
	UV	704.34	961.76	1,884.27

Source: table by FAO.org.

8.5. THE FUTURE OF FISHERIES MANAGEMENT

No other group of animals is as hard to control as a group of fish, but the technology is offering solutions to the difficulties of the sustainable fisheries management.Novel technologies at the frontier of science have a very significant and meaningful role with respect to providing assistance in the challenges of the fisheries management, from the use of Deoxyribose Nucleic Acid (DNA) in helping out in the development of the stock evaluations to the satellites combating illegal fishing.

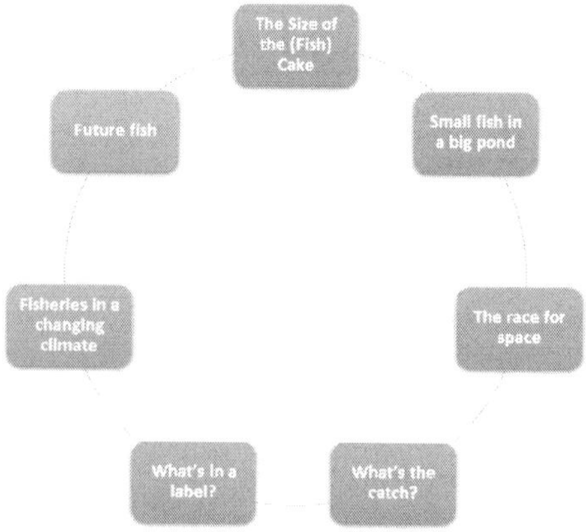

Figure 8.5. The future of fisheries management.

8.5.1. The size of the (fish) cake

In the year of 2014, the EU's fishing fleet generated near about €7.2 billion and the livelihood of several numbers of citizens of EU are relying on the sea, seafood's, and its resources. There are near about a hundred thousand vessels registered in the EU and combined, these catch near about 6.4 million tonnes of fish.

According to estimation of FAO (the Food and Agricultural Organization), all across the globe, near about 91-93 million tonnes of fish are captured with each passing year, and seafood products are among the most widely traded food commodities of the world. Its export value of near about one hundred and forty-two billion dollars (which are around €114.9) in the year 2016. Fish is the means of subsistence for billions of the people all across

Near about 80% of this capture is caught from the wild and the remainder from the aquaculture.

"If you can't measure it, you can't manage it"

Shellfish and fish are hidden from the view, but understanding what type of production is sustainable requires an approximation of the abundance of fish. Unlike terrestrial wildlife, fishery executives cannot look into the ocean and count the fish that they see.

As an alternative, they must try to approximate the size of the fish stocks as precisely as possible, which is based on the samples of the population.

The techniques that are used conventionally with origins in the management of whaling, fisheries management has largely been concentrated on the single species stock assessments.Never the less, with the increment in the computing power now readily available, evaluation of stocks is shifting towards the 'ecosystem assessments'; ecosystem assessments take into account the interconnectedness of the fish food webs and the impacts of the factors, for example, weather.

With respect to this context, the molecular biology, which is the detection of adaptive diversity (which is outer loci), that provides the unprecedented changes in order to understand the bio-complicated in the near future; the fisheries sciences can combine the evolutionary ideologies with the fisheries management in order to evaluate the potential effects of the fishing or sudden changes in climate on the resilience of fisheries stocks.

1. The beginning of the positioning systems and sonar systems, which is a technology that is adapted from the military use, and it is marked as a valuable turning point in the race to fish.
2. Fishing marks and fishing shoals, which were previously known with the help of conventional methods, were more readily available to the fishing industry, increment in the effectiveness and contributing to the decrement in the stocks of several types of fish. In the present interval of time, these kinds of technologies are also used with the help of scientists and managers, in the means of map stocks and fish habitats.
3. Sonar with multibeam can offer intricate the three-dimensional maps of the seabed, which provides assistance in permitting the managers to map the 'essential fish habitat' (EFH), for example, regions that are important for the reproduction of fish. This makes it much easier in order to identify the sensitive marine regions where fishing could harm the precious habitats.
4. More and more 'Autonomous Underwater Vehicles,' which is unmanned submarines, and that are being deployed in order to collect these kinds of data and to frequently map the regions of seabed.

8.5.2. Small fish in a big pond

Several types of species of fish shows complicated histories of life. Development and improvements in the technology that are being used in answering some of the most elusive queries in the fisheries management, for example, where do certain kinds of fish go?

Several numbers of fish migrate, some of them migrate across the oceans, while some of them migrate across both freshwater as well as salt water, such as seatrout and salmon.

In this way, in order to regulate and safeguard the fish stock at the vulnerable intervals of their lifecycle, first of all, everyone must know where they are traveling. Here, high-throughput, also known as next generation, sequencing of DNA (Deoxyribose Nucleic Acid) helps in providing the answers.

The technique permits to sequence the RNA (Ribonucleic Acid) and DNA, with higher rate and at cheap price as compared to the previous approaches; it is transforming the genomics as well as the molecular biology.

Research is now being conducted on which fish live where, which is based on the DNA in the water. as fish move through water, they shed cells, for example, scales, and with the help of measuring the presence and diversity of the cells in the water it is possible to find out the presence or passage of the animals.

Where before the fisheries managers are dependent on the expensive research vessels, with the help of updated and enhanced fishing techniques, new methodologies and techniques to the use of the 'environmental DNA' are providing the fast as well as cost effective techniques in order to understand whether, and in what abundance, first have been in the water column.

Individual fish are also being followed up in order to provide the insights into their migration and movements. Pictures of the fish migrations have previously been built from the models of occurrence or frequency of capture and left large voids in the knowledge for several numbers of species.

The development of the pop-up satellite archival tag (PSAT) technology that tags inserted or attached to fish, which provides assistance in order to improve the acquisition of this information and revolutionizing investigations into the distribution of the migratory species.

8.5.3. The race for space

Along with, providing help in where fish live and move from and to, technology is helping in mapping and monitoring the fishing activity. Monitoring where, when, and how, fishing fleets are functioning has conventionally depended on the self-reporting and fisheries monitoring with the help of ships, port inspection, and aerial assets. But also, once over the horizon, surveillance of fishing vessels is very hard task and expensive as well.

In the present interval of time, satellites are frequently being utilized in order to provide assistance in monitoring the fishing activities. In Europe, all vessels over 12 meters in length are needed in order to operate a Vessel Monitoring System (VMS). These kinds of systems send the location speed and the direction of a fishing vessel with the help of satellite in order to regulate the authorities.

The technology is also available to (smaller) inshore fleets, which functions closer to land, with the help of network of mobile phone. This use of the technology enormously decreases the compliance cost and offers manager with a much better understanding of the distribution of the fishing activity. This is particularly helpful in making sure that the specific types of habitats in 'MPAs' are not harmed.

The NASA-NOAA Visible Infrared Imaging Radiometer Suite (VIIRS) can identify lit fishing boats at night and this technology is being used in order to find out the fishing in the further reaches of our oceans, for example, the regions that are far from routine monitoring, and sometimes outside states fisheries controls.

8.5.4. What's the catch?

In addition to the specific kind of species, other marine fish animals may be caught; if discarded, this practice is totally waste of time, and if safeguarded species are caught it can harm the conservation efforts. Conventionally, the fisheries observers would be deployed in order to monitor the activities, but with respect to the cost as well as conditions, only constrained observations can be made.

With the help of execution of the high-quality camera, low-cost, and the devices to store the data, nets, and fishing operations can be observed or monitored in order to provide the precise approximation of the by catch. The vast data is generated with the help of which these systems can be analyzed with the help of the artificial intelligence.

In which the computers are trained in order to find out the specific kinds of patterns in the catches. This data can be then informing the management of fishing gear design or fish allocation in certain regions, so as in order to minimize the unwanted by catch and safeguard the environment.

8.5.5. What's in a label?

Industry of fisheries is quite a bug business and the supply chain are complex and the chances of fraud exists. In the year 2010, according to the study of

EU, they found that, in near about 40% cases of the fish consumers thought they were buying was mislabeled and was substituted for the other species.

The profits to do so are fluctuated, but every so often, cheaper fish may be used and labeled in order to command a higher price. From the study done in the year 2010, subsequent research has disclosed that the incidence of mislabeling may have decreased, as an outcome of the ability to routinely sample fish, with the help of RNA and DNA sequencing to check the providence.

Increment in the global as well as EU requirements in order to label fish from 'net to late' means that the consumers and buyers can make informed choices as to the fish they eat and or serve. In this way, great enhancements with respect to the traceability have been made, providing the incentives or benefits for environmentally sensitive fisheries in a market which requires a premium for these kinds of products.

Food labeling, such as the Marine Stewardship Council (MSC) certification which is recognized all across the globe, which further enables to make choices about the fish that they eat.

8.5.6. Fisheries in a changing climate

All across the globe, temperature of sea has increased over the time period of several years and as widely-documented, every so often northward. Shifts in the ranges of the several numbers of species of fish that are being associated to such altering environmental conditions.

In the context to this, techniques, and technologies, such as broad-spectrum satellites, are providing assistance to the scientists in order to monitor the allocation of the algae and plankton, which is the base of several numbers of fisheries food chain. As the global conditions change and fish stock respond, fisheries management will require to adapt; technological tools are the forefront of this challenge.

8.5.7. Future of fish

In the context of Europe, an increment in the number of fish stocks are controlled in a sustainable manner. Never the less, in spite of meaningful enhancements more than 40% of the fish stocks in the North Sea, that is the most productive sea in all across the Europe, are overfished. The situation is much worse, and the challenges are must vaster, with respect to the Mediterranean Sea.

The opportunities in order to improve both the economic as well as environmental results of sustainably managed fish stocks are meaningful and the scientists approximate that if it controlled sustainably, stocks could harvest an additional 1.45 tonnes of fish per year within the time span of next one decade. In order to accomplish this, the further adoption and the development of the technology is very important.

Fish stocks are the largest wild food resource and, if they are carefully organized, they can provide employment and food within environmental limits.

In addition to this, further access to, investment in and integration of the technology into the fisheries management sector offer a chance in order to accomplish the vision of Blue Growth of EU.

8.6. THE FUTURE FOR FISHERIES

Generally, fisheries are perceived as local affairs requiring, with respect to the scientific inputs, annual reassessments of the catch quota which is species-specific.

Most of the scientists that deals with the fisheries are employed with the help of regulatory organizations in order to generate the quotas, which ideally should make fisheries sustainable and profitable, contributors to the employment, with the help of international trade, to global food security.

This perception of fisheries as local and species-specific, managed to directly benefit the fishers themselves is conducive neither to global anticipations nor the collaborative development of long-term scenarios.

Certainly, latest amounts of this kind, expect those of the United Nations Food and Agriculture Organization (FAO), tend to be more self-conscious and surrounded in irony, maybe a suitable response to 19th century notions of inexhaustibility.

The time period of last ten years, have established that the fisheries must be viewed as the elements of a global enterprise, on its way to undermine its supporting ecosystems.

These kinds of developments take place against a backdrop of fishing industry lobbyists arguing that the governments drop some rules that produces trouble and the economists are assuming that free markets produce inexhaustibility.

The aquaculture sector provides to feed the world with farmed fish. On the other hand, building more coastal feedlots, where the carnivores, for example, salmon, and tuna are fed by the other fish, the aquatic equivalent of robbing Peter to pay Paul.

1. The time has come to look at the future of the fisheries with the help of;
2. Identification as well as exploration of the basic trends; and
3. Development and exploration (with or without the help of computer simulation) of possible futures.

The research community of the fisheries depends on the broad-based analysis, on a data set now shown to be severely biased. First-order correction recommends that in spite of increasing, as previously reported, fisheries landing is decreasing by near about five hundred thousand metric tons per year from a peak of 80 to 85 million tons in the time period of late 1980s.

Because overfishing and decrement in the habitat are likely to endure, extrapolation can be well thought out as well.

This correction, never the less, does not consider discarded "by-catch" (near about 30% of the landings all across the world) only one component of the illegal, unreported, or unregulated (IUU) catches that presently became part of the international fisheries research agenda.

The geographic and the depth expansion of the fisheries is very easy to extrapolate. Over the time span of near about five decades, fisheries targeting benthic and bentho-pelagic organisms have covered the shelves that surrounded by the continents and islands down to two hundred meters, with an increment in the inroads below a thousand meters. On the other hand, fisheries targeting the oceanic tuna, billfishes, and their relatives that covers the world ocean by the early.

Extrapolating the bottom fisheries trends to the year of 2050 is quite straightforward. With the help of satellite positioning and seafloor-imaging systems, humans will deplete the canyons, deep slopes, seamounts, and the deep-ocean ridges of the local accumulations of the judiciously renamed bottom fishes.

For example, orange roughy (formerly known as "slimeheads"), Chilean seabass (usually IUU-caught Patagonian toothfish), and hagfish (caught for their "eel skins," and here anticipated to become a delicacy in the trendy restaurants, freshly knotted and sautéed in their own slime), the abyssal tripod fishes being the only group that seems safe so far.

Figure 8.6. An illustration of conventional fishing.

Source: Image by national Park Service.

The radical trend change required to turn 20% of the shallowest hundred meter of the world ocean into the marine reserves by the year of 2020, that is returning to the 1970s state.

One effect may be to elevate the human consumption of small pelagic (herrings, anchovies, mackerels, or sardines, for example, the Peruvian anchoveta), now mostly turned into the fish meal for agriculture (in order to grow the pigs and chickens, and for use as fertilizer) and aquaculture.

Never the less, the anticipation is better embedded into the scenarios, which is the set of coherent, plausible stories designed in order to address complicated questions about an uncertain future.

Scenario analysis is more specifically important for the fisheries sector, which, never the less a major provider of food and jobs in several numbers of poorer nations, is quite small as compared to the economy of richer nations and in this way, "downstream" from most of the policy decisions.

Pending the detailed analysis of the coastal as well as marine scenarios with the help of Millennium Ecosystem Assessment, four scenarios has been developed with the help of United Nations Environment Programme (UNEP), in order to investigate the future of marine fisheries. For each and every case, there is a summarized results of regional simulation models explicitly accounting for interspecies feeding interactions, within a range of ecosystem kinds and fisheries.

1. Markets first, where market considerations make environmental policy. This may point at the gradual elimination of the subsidies drives the overfishing. Putting markets first may also suggest

the suppression of IUU fishing (which is consisting of flags of convenience), which distorts the economic rationality as insider trading or fraudulent accounting does.

Market first, by overcoming subsidies, could also take into the direction to the decommissioning of fuel-guzzling distant water fleets (more specifically large trawlers), and perhaps takes into the direction to a resurgence of small-scale fleets deploying fixed gears that are energy efficient.

This case permits for spontaneous emergence of quasi-marine reserves (that is, areas that are not economically fishable, more specifically offshore) and in this way, it can decrease the impact on the biodiversity. Never the less, groupers, bluefin tuna, and other taxa (which is consisting of invertebrates) which are very costly, would remain under the pressure.

2. Security first, where conflicts along with the inequality takes into the direction to a strong socio-economic limit between rich and poor. This case, never the less depicts some suppression of IUU fishing, would continue "fishing down marine food webs," which is consisting in the High Arctic, and subsidization of fleets to their logical ends of rich nations, which is consisting of the collapse of the conventional fish stocks.

This shows the development of the alternative fisheries targeting the jellyfish as well as other zooplankton (more specifically, krill) for direct human consumption and as feed for farmed fish. Generally, this scenario emphasizing present ("south to north") patterns of trading, would largely eliminate the fish from the markets of nations which will be still in development in the year 2050.This scenario would also elevate the export of polluting technologies to poorer nations, particularly coastal aquaculture and/or fertilizers of the open sea. This would have negative influence or effects on the remaining marine fisheries in the host nations, with the help of harmful algal blooms, invasive species, and diseases.

3. Policy first, where a range of actions is undertaken with the help of governments in order to balance the social equity and the environmental concerns.

This is illustrated with the help of recent Pew Oceans Commission Report, which for the United States, proposes a new Department of Ocean and regional Ecosystem Councils, and a reform of the Fisheries Management Councils, now they run with the help of self-interested agencies.Identical regulatory reforms, that is coordinated between the nations, integrated with the marine reserve networks, huge decrement in the fishing effort, more specifically gears that destroy the bottom habitat and produces large "by catch," and abatement of the coastal pollution, can bring fisheries back from the brink and decrease the threat of extermination with respect to several numbers of species.

This scenario corresponds to the simulations where rest is maximized subject in the context of biodiversity constraints.

4. Sustainability first necessitates a value system change, favoring the environmental sustainability. This scenario, which shows the ratification of and adherence to the international fishery management agreements of the government and bottom-up governance of the local resources, would consist the creating of the network of marine reserves and carefully monitoring and rebuilding a number of major stocks.

The reason behind is this that the high biomass provides the best protection with respect to the overestimates of catch quotas and environmental changes.

These scenarios explain that, what might occur, not what will come to pass. But still, they can be utilized in order to consider what human wants for their future. It has been noted that, never the less, several numbers of the fisheries that are being investigated, for example, in the regions of North Atlantic or the Gulf of Thailand, currently optimize nothing of benefit or advantage to the society: not rent, not even gross catches (and in this way, long-term food and the employment security).

8.7. CONCLUSION

Fisheries, especially the market-based fisheries, have been heavily impacted by the increase in the rate of human intervention. The traditional fisheries have tried to adopt the methodology of the modern-day fishery management. Although, the modern-day fishery management methodologies are very effective in handling the SSF. But when it comes to handling market-based fisheries, it becomes difficult to dwell and grow with the same rate as that of conventional fisheries.Also, the future of the market-based fisheries has been heavily impacted with the present-day human activities and intervention. Many instances of overfishing and oil spills have heavily impacted the future of fisheries, decreasing the life of the fishes and spoiling their entire ecosystem.

REFERENCES

1. B., Hanich (2014). *Fish for the Future: Fisheries Development and Food Security for Kiribati in an Era of Global Climate Change | WorldFish Publications.*[online] Worldfishcenter.org. Available at: https://www.worldfishcenter.org/content/fish-future-fisheries-development-and-food-security-kiribati-era-global-climate-change-0 (accessed on 8 March 2020).
2. Daniel, P., Jackie, A., Elena, M. B., Villy, C., Peter, T., & Reg, W., (2003). *The Future for Fisheries.* [eBook] science mag. Available at: https://www.researchgate.net/publication/235417771_The_Future_for_Fisheries (accessed on 8 March 2020).
3. FAO (2008). *Present and Future Markets for Fish and Fish Products from Small-Scale Fisheries -Case Studies from Asia, Africa, and Latin America.* [eBook] Rome: Food and agriculture organization of the united nations. Available at: http://www.fao.org/3/a-i0230e.pdf (accessed on 8 March 2020).
4. Leadbitter, D., (n.d.). Market-Based Mechanisms-Improving Fisheries Management? *Seafood Eco-labeling*, [online] pp. 187–206. Available at: https://www.researchgate.net/publication/228002369_Market-Based_Mechanisms_-_Improving_Fisheries_Management (accessed on 8 March 2020).
5. Robert Clark, (2018). *The Future of Fisheries Management | Government Europa.* [online] Government Europa. Available at: https://www.governmenteuropa.eu/future-fisheries-management/85633/ (accessed on 8 March 2020).
6. Secretary of State for Environment, (2018). *Sustainable Fisheries for Future Generations.* [eBook] Crown. Available at: https://assets.publishing.service.gov.uk/government/uploads/system/uploads/attachment_data/file/722074/fisheries-wp-consult-document.pdf (accessed on 8 March 2020).
7. Solway Firth Partnership, (2019). *Extended Closing Date for Future of Fisheries Management in Scotland Consultation-Solway Firth Partnership.* [online] Solway Firth Partnership. Available at: https://www.solwayfirthpartnership.co.uk/extended-closing-date-for-future-of-fisheries-management-in-scotland-consultation/ (accessed on 8 March 2020).
8. Zeller, D., & Pauly, D., (2019). Viewpoint: Back to the future for

fisheries, where will we choose to go? Global Sustainability, [online] Vol. 2. Available at: https://www.cambridge.org/core/journals/global-sustainability/article/viewpoint-back-to-the-future-for-fisheries-where-will-we-choose-to-go/426C9BB9369552CC81A1A10E9557 3F53 (accessed on 8 March 2020).

INDEX

C

D

E

F

G

H

I

K

L

M

N

O

P